PRAISE FOR
THE MILKY WAY

"As a character, the Milky Way is a cross between a Greek goddess and GLaDOS, the artificially superintelligent computer system from the Portal video-game series. She gossips about other galaxies, teaches us about her past, and imparts a primer on astrophysics, all the while relishing every opportunity to throw shade on humankind's egocentrism and closed-mindedness."

—***Scientific American***

"Astrophysicist McTier delivers in her debut a delightful report on the Milky Way's inner workings... McTier writes that her goal is to help people 'understand how ephemeral [our] existence is.' She succeeds smashingly. The result is truly stellar."

— *Publishers Weekly*, starred review

"Creative, humorous, and enormously entertaining... As with any translation from another tongue, readers may marvel at the role of the translator in creating a book that is both informative and truly inspirational. Here, it's clear Dr. McTier has harnessed the sense of marvel she felt as a child, when she imagined the sun and moon as celestial parents who watched over her and talked to her on a regular basis. That childlike wonder, combined with her expertise in mythology and astronomy, makes her the perfect human to assist in telling this story." — ***BookPage*, starred review**

"It's about time we heard the story of the Milky Way in its own words. The good news is that our galaxy is not only ancient and majestic; it's also whimsical, amusing, and downright chatty. Moiya McTier's book is an entertaining introduction to some of the most profound features of our astrophysical neighborhood."

—Sean Carroll, *New York Times* bestselling author of *Something Deeply Hidden*

"If you want to learn about the Milky Way, who better to go to than the source? Well, up until now, the Galaxy hasn't been talking—but all of that has changed! Turns out, the Milky Way has a sense of humor, an attitude, and, frankly, isn't super impressed with us as of late. If you're looking for a fun and unique way to learn about astrophysics—this is the book for you!"

—Kelly Weinersmith, *New York Times* bestselling author of *Soonish*

"A direct, fun, and charming mix of the science, folklore, and history of our Milky Way galaxy. And since that galaxy is technically composed at least in part by ME, I cannot help but take some of the credit."

—Ryan North, *New York Times* bestselling author of *How to Invent Everything*

"Moiya McTier gives us an exciting romp through the universe from the perspective of a most unexpected guide: our local sentient collection of stars, gas, dark matter, planets, and its wayward humans. What an exciting way to learn about everything in the universe, from its earliest moments to star births and deaths. Only here will you learn what the Milky Way thinks of its neighbors. McTier invents the genre of cosmic gossip—what fun it is!"

—Chanda Prescod-Weinstein, author of *The Disordered Cosmos: A Journey into Dark Matter, Spacetime, and Dreams Deferred*

"Moiya McTier tells the story of our galaxy in a whole new way in her delightful new book."

—Space.com

"Endearing and entertaining... *The Milky Way* reminded me to look up and appreciate my home in the universe, just like its narrator wanted."

—ScienceNews.org

"[A] one-of-a-kind look at our galaxy... Educational, informative, and original, this will leave readers eagerly anticipating McTier's next book."

— *Booklist*

"McTier engagingly narrates her delightful debut... [An] educational and fun take on the stars."

— *Library Journal*

"McTier sprinkles humor throughout her whimsical look at the cosmos...The author clearly knows her subject and delivers enough fascinating information to keep the pages turning."

— ***Kirkus Reviews***

"McTier's sharp wit and sharper intellect strike the perfect tone for this breezy take on the history of our galaxy. Truly the biggest tell-all story in the universe!"

—Paul M. Sutter, PhD, astrophysicist and author of *Your Place in the Universe*

"Brilliantly blending astrophysics and mythology, McTier has crafted an out-of-this-world work of genius. *The Milky Way* is a remarkably clever, eye-opening entry into the astrophysics cannon that radically changes our perspective on space and our place in the vast cosmos. As entertaining as it is informative, this book is an essential read for earth dwellers who want a better understanding of our galactic home."

—Stephon Alexander, author of *Fear of a Black Universe*

"A deliciously hilarious/irreverent and irresistible romp shimmering with astrophysics facts and cutting-edge observations. A first 'person' perspective that only an astrophysicist can provide. McTier's humor and keen eye for detail pens the autobiography that our home galaxy deserves!"

—Brian Keating, Chancellor's Distinguished Professor of Physics at UC San Diego and author of *Losing the Nobel Prize* and *Into the Impossible: Think Like a Nobel Prize Winner*

"We usually imagine the Milky Way as just a space full of dust and darkness with no other lifeforms in it, but in this book, it is represented as something much more lively and exciting... Go on this journey to the depths of our home to learn more about it, and to reflect on ourselves!"

—**Astrobites**

"The book is a good introduction into the past, present, and future of our galaxy and our universe. There is no single ideal approach for communicating astrophysics, or other science topics, to all readers, and alternative approaches like this may attract those turned off by a more conventional approach."

—**The Space Review**

THE MILKY WAY

THE MILKY WAY

AN AUTOBIOGRAPHY OF OUR GALAXY

Via

MOIYA McTIER

ILLUSTRATED BY ANNAMARIE SALAI

NEW YORK BOSTON

Grand Central Publishing
Hachette Book Group
1290 Avenue of the Americas, New York, NY 10104
grandcentralpublishing.com
twitter.com/grandcentralpub

Originally published in hardcover and ebook by Grand Central Publishing in August 2022

First trade paperback edition: August 2023

Grand Central Publishing is a division of Hachette Book Group, Inc. The Grand Central Publishing name and logo is a trademark of Hachette Book Group, Inc.

The publisher is not responsible for websites (or their content) that are not owned by the publisher.

The Hachette Speakers Bureau provides a wide range of authors for speaking events. To find out more, go to hachettespeakersbureau.com or email HachetteSpeakers@hbgusa.com.

Grand Central Publishing books may be purchased in bulk for business, educational, or promotional use. For information, please contact your local bookseller or the Hachette Book Group Special Markets Department at special.markets@hbgusa.com.

Library of Congress Cataloging-in-Publication Data

Names: McTier, Moiya, author.
Title: The Milky Way : an autobiography of our galaxy / via Moiya McTier.
Description: First edition. | New York, NY : Grand Central Publishing, 2022. | Includes bibliographical references and index.
Identifiers: LCCN 2022004461 | ISBN 9781538754153 (hardcover) | ISBN 9781538754177 (ebook)
Subjects: LCSH: Cosmology--Popular works. | Milky Way--Popular works.
Classification: LCC QB857.7 .M48 2022 | DDC 523.1/13--dc23/eng20220421
LC record available at https://lccn.loc.gov/2022004461

ISBNs: 9781538754160 (trade paperback), 9781538754177 (ebook)

Printed in the United States of America

LSC-C

Printing 1, 2023

To everyone who's ever been made to
feel that they're not "sciencey enough,"
whatever that means.

CONTENTS

Foreword from Moiya		*xv*
CHAPTER ONE:	I Am the Milky Way	1
CHAPTER TWO:	My Names	13
CHAPTER THREE:	Early Years	18
CHAPTER FOUR:	Creation	37
CHAPTER FIVE:	Hometown	42
CHAPTER SIX:	Body	67
CHAPTER SEVEN:	Modern Myths	88
CHAPTER EIGHT:	Growing Pains	96
CHAPTER NINE:	Inner Turmoil	116
CHAPTER TEN:	Afterlife	136
CHAPTER ELEVEN:	Constellations	144
CHAPTER TWELVE:	Crush	151

CHAPTER THIRTEEN: Death 169

CHAPTER FOURTEEN: Doomsday 188

CHAPTER FIFTEEN: Secrets 195

Acknowledgments 217

Notes 221

Index 235

Reading Group Guide 245

About the Author 251

FOREWORD FROM MOIYA

"I HAVE LOVED THE STARS too fondly to be fearful of the night."

This last line from Sarah Williams's poem "The Old Astronomer to His Pupil" has often been a sort of mantra for me. And not just because it makes me sound like a spooky Victorian recluse.

I don't remember how, but as a young kid, I got it into my head that the sun and moon were my celestial parents. I imagined that they watched over me, and I actually talked to them, told them about what I was learning in school and what my friends were like (because, as I was surprised to learn, those friends *didn't* talk to the moon and sun, so *someone* had to tell our celestial mom and dad what was up). When my earth parents started having arguments at night, I cried to my celestial mom. And when my birth dad stopped showing up for scheduled home exchanges,[i] my little kid mind decided to blame the sun, too. To this day, I don't like LA because it's too sunny.

My earth mom fell in love again and we moved from our small

---

i Don't worry, we've worked it out since.

Pittsburgh apartment to the strangest place I could imagine: a log cabin in the middle of the woods without running water, so close to the West Virginia border that I had to cross state lines to reach the nearest bookstore. The forest was the best playground an only child[ii] could ask for, a space to invent epic quests, hunt for faerie rings, or find the perfect branch to use as a fighting staff in mock battles with my new earth dad. But the community surrounding that forest, full of people who had only ever seen a Black person on TV, probably wasn't the hometown I would have picked for myself if my mom had asked.

For that reason and so many more (you try getting your period at ten years old when you don't have a working shower in your home), I sought comfort from the moon well into my adolescence. I developed a great love for the nighttime, the time of quiet, secrets, and peace. Declaring myself a creature of the night helped cement my desired place as the Weird Kid, as if being the smartest and also the blackest person at my small, rural school didn't already make me stand out enough. That's not an empty flex; I was voted most unique and was easily the valedictorian of my class *after* skipping sophomore year. People still said I only got into college because of affirmative action.

Don't get me wrong, most people I interacted with were very kind, and I'm grateful for the experiences had and connections made that let me empathize with a part of the country that rightfully feels ignored by the very class of intellectual elites I worked

---

ii I do have some half brothers on my birth dad's side, but obviously I didn't grow up with them because, well, see previous note.

my ass off to break into. I learned valuable lessons in coal country, like how to chop firewood, do a deep-conditioning treatment with nothing but a bucket of water and a cup, and look past obvious differences to find common ground. But I also learned early on that my life would be better if I got myself out of there ASAP. Lucky for me, Harvard admissions officers are much fonder of bizarre, brilliant, Black girls than a lot of miners' sons were.

Even though I always felt most comfortable at night and lived in a place with a beautiful view of the stars, I was never interested in the academic study of space before college. I merely loved the celestial aesthetic. But it didn't take me long to fall in love with the logical, data-driven nature of astronomy. The summer after sophomore year, I did a research internship where I spent hours analyzing five-dimensional data cubes to measure properties of a distant star-forming galaxy that I nicknamed Rosie. Falling deeper into astrophysics felt like learning how to talk to space in a whole new way, one that let me *listen* a little bit more to what the universe was saying instead of making up responses in my head. I was learning the language of gravity, cosmic rays, and nuclear fusion. With my new dictionary in hand, I set out to research as many different aspects of space as possible: star formation, the cosmic microwave background, X-rays from distant quasars, exoplanet characterization, stellar dynamics, and the chemical evolution of galaxies.

At the same time, following my love of mythology, I was learning about the stories that cultures used as devices to entertain, educate, and explain. Fairy tales to pass a night by the fire, fables to share a community's values with the next generation, and myths to make sense of the world around them. I realized that, like my unusual mix of backgrounds, science and myth weren't as contradictory as they seemed on the surface. Both are tools that we humans use to understand how we fit in with the rest of the universe. And after spending almost ten years studying the physics of space, five of them in a PhD program that inspired three stress tattoos and multiple rounds of therapy, my perspective on everything has widened in the most illuminating way. I feel more connected to people and nature, and more comfortable with my place among all of it.

Astronauts feel this same shift in perspective when they view Earth from orbit, because when you're in space, you can't see the imaginary borders that divide us. You see how fragile the complex, interconnected ecosystem we call home really is, and our petty human squabbles seem small and unnecessary. Philosopher Frank White called this life-changing cognitive shift the overview effect, and I've always thought Earth would be a much nicer place for all of us to live if we each got to experience just a little bit of it.

Realistically, we won't all get there by visiting space. Some people get to this same point through faith or meditation or drugs. I got there through science, by spending an inordinate amount of time picturing Earth, our solar system, and the Milky Way as their own small parts of the grander whole. Okay, maybe there were some drugs, too, but it was mostly the way the science mixed with my gentle artist's soul.

Now that I know how to speak its language, I'm more enamored of nighttime than ever. That's why I was honored when the Milky Way itself picked me to relay its story. I hope that by the end, you've grown so fond of the stars and the galaxy that made them that you, too, start to hear what the night has to say.

CHAPTER ONE

I AM THE MILKY WAY

TAKE A LOOK AROUND YOU, human. What do you see?

Actually, don't answer that. Why would I bother listening to you when I know you'll get it wrong? You'll start naming objects and places, but that chair you're sitting in isn't just a chair. That book you're holding isn't just a book. Even the planet your kind is on the brink of ruining isn't just a planet. They're all *me*.

Everything you've ever seen or touched is a part of me. Yes, even you, you vain, filthy animal.

I made it all. Not intentionally, of course. I have no need for chairs, and I really couldn't have cared less about whether or not one of my worlds produced life, especially in a form that was so picky about where it sat. You humans just *appeared* one millennium, and then it took another several thousand years for me to actually notice you. I guess, in some ways, I'm glad that I did. (But if anyone else ever asks me, I will absolutely deny feeling any sort of affection for your fleshy species.)

Before we get too far along, allow me to introduce myself. I am the Milky Way, home to more than one hundred billion stars (and yet you still think yours is special enough to have its own name) and the fifty undecillion[1] (that's five followed by thirty-seven zeros) tons of gas between them. I am space; I am made of space; and I am surrounded by space. I am the greatest galaxy who has ever lived.

If you have even a portion of the requisite curiosity needed to engage with this volume, you might be thinking to yourself, "How can the Milky Way talk?" Well, with your short lives, you certainly don't have enough time for me to teach you everything there is to know about theoretical physics and schools of consciousness, but I can tell you a theory or two that might answer your question.

Some of your human physicists predicted what they considered to be an absurd consequence of your second law of thermodynamics, which says that the entropy of a closed system always increases. In other words, the universe as a whole should always be trending towards chaos. But how can that be true if our universe appears to be so organized? One possible explanation, which your physicists have since learned is wrong (this will become a trend), is that our universe as we see it is simply a very fortunate but extremely random distribution of matter. The extreme consequence of that explanation was that as entropy increased and more random fluctuations appeared, some of that matter should take the form of human brains,[2] or at least a similar network of thought cells. Your physicists thought the idea was ridiculous, but you'll soon see that there are plenty of seemingly random fluctuations in the universe. And if matter can combine to form brain-like systems on your little planet, why shouldn't it do the same everywhere else?

Separately, your philosophers have postulated that consciousness isn't a quality inherent to humans, or even living animals. According to them, consciousness, or sentience or awareness or whatever you want to call it, is the result of how a system *functions*, not a consequence of what it's made of. Some of your philosophers are even starting to believe that consciousness is an inherent quality of the universe, something that every amount of matter possesses in different quantities. In other words, I can think and communicate even though I don't have what you would consider a brain. So if you're imagining I'm anything like one of you, cease immediately! It's insulting, and that human-centric mindset will just make it harder for you to understand all that I am going to deign to teach you.

If your question was more like, "How can the Milky Way talk to *me*," well, it's not like human language is that hard to learn. You're such simple creatures.

Now that the obvious questions are out of the way, you're probably wondering why I—the greatest galaxy ever, who never even wanted humans to exist in the first place—have chosen to communicate with you.

Whether I like it or not, our lives are intertwined. My existence is, of course, much more important to you than yours is to me, but over time your kind has demonstrated that you aren't completely useless. (You'll have to forgive me if I don't always phrase things in the most pleasant way; the concept of niceties in the manner that you deploy them is fairly new to me. Also, you'll be dead soon, so why should I care if I hurt your precious feelings?)

You see, as far as I can tell, I'm more than thirteen billion of your Earth years old. The story of my glorious birth will come later, but all you need to know right now is that I'm nearly as old as time itself. To use a comparison your kind seems to be fond of—even though it's not even close to an adequate description of my age—I am literally older than dirt. I was alive when the individual atoms that make up your dirt were created billions of light-years away from where they are now. For most of that time, I've been so bored and—though it may not look like it to you—lonely.

If you've heard anything about me at all, you probably think that my life must be so glamorous and full of important, gratifying tasks. Creating all those stars, building all those planets, molding the very essence of the universe according to my will like clay . . . yeah, it was the ultimate thrill. For a few billion years.

There are only so many new perfect combinations of stars, planets, and moons that a galaxy can forge, so I started making imperfect ones. I experimented until I made something that was kind of a star and kind of a planet, but ultimately failed at being both.[3] I flung black holes at each other until I became numb to the ripples they produced. I built planets on orbits that I knew would result in their either spiraling into their stars or getting flung out of their systems. Hot Jupiters[4] that orbit mysteriously close to their stars? Yeah, that was just a casual experiment, and now they're everywhere. You're welcome, astronomers.

You likely can't relate, but even being the best at something gets

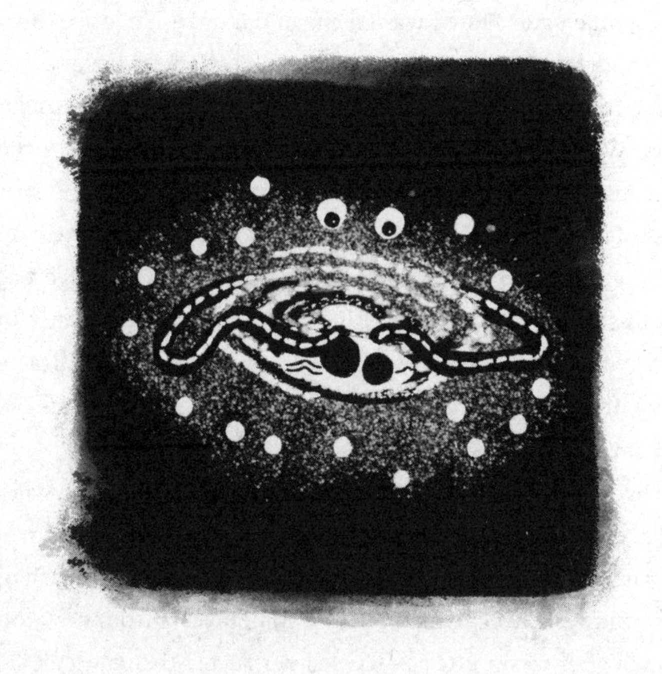

old after a while. So, when the beautiful chaos I'd created stopped exciting me, I put it all on autopilot. That's why I became much less active nine billion years ago. Your astronomers have noticed that I slowed down my star production back then, but they all chalked it up to a decrease in available star-forming gas. They're technically not wrong, but did they ever think to ask me why I lost so much gas? How I was feeling at the time? No, none of you ever think to ask me anything anymore. That's the problem.

You might be wondering what I was doing in those nine billion years. Well, while what I do in my sleep is orders of magnitude more impressive than anything you could ever accomplish, I spent most of my time thinking. You know, reflecting on past deeds and reveling in my triumphs. I passed the occasional message back and forth with other galaxies in my neighborhood, mostly the dwarf satellites who hang around because they're just so attracted to me. Literally. It's a gravity thing. I've grown a bit fond of some of them.

That might not seem like much activity to fill nine billion years, but you must remember that our lives don't operate on the same timescales. I've already lived for more than ten billion years and I'll still be living at least a trillion years from now, long enough after your puny sun has self-destructed that an exact date is meaningless. It would be generous of me to compare your life span to a blink of my eye, except I don't actually have eyes. You can call someone on the other side of your world and talk to them immediately with the help of signals traveling at the speed of light. It takes me more than twenty-five thousand years to send a light message to my nearest neighbor. Taking a million years to think about that one time I said, "You too," when another galaxy told me to enjoy my supernova? That's nothing.

I'm getting carried away, which you'll realize happens often. My point is that I was steeping in my own thoughts for literal eons until you humans popped up about two hundred thousand years ago.

It was...astounding how much you didn't understand. And I wouldn't say you've come much closer to solving the deepest mysteries of the universe, but at least back then humans knew the most important thing: that I'm incredible.

Through your stories, you taught your children to look up at me when they lost their way. It took you ages to stop chasing all those four-legged creatures—some of you still do—but eventually you figured out that you could track my motion to determine the best time to plant your crops. And I saved thousands of lives once you learned you could use me to predict oncoming disasters. That wasn't just your ancestors attempting a kind of magic; it was their knowing that my movement aligned with cyclical events in nature like regular floods[5] or insect swarms, even if they did often end up explaining away such events with magic or religion.

Your stories made me feel loved and needed and, perhaps for the first time in my long existence, more helpful than I was ruinous. Every galaxy should feel so lucky as to know it has positively affected the universe. Well, for other galaxies, it's luck. For me, it's just raw beneficent talent.

It's not that I craved your attention or needed a group of people to worship the ground that I don't walk on. I wasn't just waiting for ten billion years for you to come around and stroke my ego.

But once you did, it was comforting to know that I could help you along. So much of what I do is destroy.

Then, in what felt like no time at all, that feeling dissipated. It started in the 1300s when you made the first mechanical clocks, and it only became worse when you invented telescopes three hundred years later and finally saw me in more detail. Once you could keep your own time and realized I wasn't merely a celestial reflection of divine will, most of you assumed you didn't need me anymore. You stopped looking up, stopped telling my stories, stopped letting me guide you. At first, I thought it was just a phase, that you were lost and would come back to me when you were ready. I've gone through enough phases of my own to afford you a brief neglectful period. Patience, after all, is one of my best qualities.

Though, in the interest of transparency—I hear that's how one builds trust on Earth, no?—I did briefly, only for like fifty years or something, consider asking your sun to throw out a flare that would wipe out all of your electronics so you would depend on me again. But you know how kids are. Just because you create them doesn't mean they'll do whatever you ask. So I graciously abandoned my murderous plan.

Then I remembered—because wisdom is another of my best qualities—that several hundred years is actually a long time for humans. Your silence wasn't just a brief distraction; entire generations had passed without bothering to think about me.

In some ways, I felt better realizing that it's not specifically *your* fault that your kind stopped caring about me. Your world is no longer set up to appreciate my splendor. It hasn't been since well before you were born. In the last one hundred years, your human cities have become blinding beacons of light that your distant ancestors never would have imagined. The electricity you all value so much has stolen something precious from nearly 80 percent of you: an unobstructed view of my gorgeous body.[6] And that's just the light pollution. The tiny smog particles that you've been overproducing since you started your little industrialization project in the 1700s aren't merely damaging your lungs and trapping heat in your planet's atmosphere. More importantly, they're blocking my light from reaching Earth's surface. There are humans alive right now who have only ever seen a handful of my stars, which is a tragedy! And I am as much a victim in this as all of you for being rendered basically invisible.

If you're an astute reader—and your choice to read this book does imply some advanced cognitive ability—then you might be wondering why I'm not satisfied by simply aiding astronomers in their research. The sad fact is that there are only about ten thousand human astronomers total, out of nearly eight billion humans. They do excellent work—honestly, it's amazing what they've been able to learn without leaving your tiny little rock—but the typical astronomy paper gets read by at most twenty other people. And those people already know most of what's in the paper anyway, so helping astronomers does little for your planet's ignorant masses.

Also, it's more entertaining to watch your astronomers struggle through the learning. When they become extra frustrated, many of them start chewing frantically on their nails, and it's just too darling to forgo by giving them the answers.

I realized that either I could remain bitter and sullen about the fact that most humans have forgotten about me, or I could do something to change it. And although I don't actually have an ass that I could get up off, to employ one of your crass expressions, I chose the latter.

The issue is that too many of you don't know enough about me to understand how I can assist you. You literally *live in me*, but most of you don't even know what I look like, let alone what I'm made of or how I move. And it's probably asking too much to expect that you would learn those things on your own. It's *definitely* asking too much of your astronomers to expect that they could effectively teach their fellow humans what they've learned. So, alas, the responsibility falls to me. Lucky for you, I'm willing and more than able to do you this service.

So here I am, introducing myself to you officially for the first time. I am the Milky Way, the galaxy whom you probably enjoyed staring at when you were young—the human children, at least, have retained enough of a sense of wonder to let me into their lives—but promptly forgot about as soon as you hit puberty and decided you had more important things to do.

I've kept your kind safe and entertained for millennia, and I'll continue doing so by telling you my story. You have a word for when a person writes about their own life: autobiography. That's what this book is. I'll tell you how I was born and where I grew up. I'll talk about my deepest shame and how I instigated the greatest love story in the universe. I'll even reveal my feelings about my—and by extension your, *if* your kind survives that long—impending death. And if my story moves you to share it with your fellow humans and maybe make up some tales of your own, then I shall consider it a triumph.

Based on what I've seen, your world isn't likely to backslide into antiquity anytime soon. Light pollution won't go away completely, and your species' days of building stone henges to track time are over. I can't guide you in the same way that I did your ancestors, but allow me to explain how you, an average modern human, can benefit from both space research and personally knowing more about the galaxy you should call home.

Take for instance that piece of technology glued to all your hands. Even I can see how much you love your cell phones, and we've already covered that I don't technically have eyes. You use them to communicate with each other, keep track of your appointments, navigate your world, and take your—ugh—*selfies*. Honestly, you use them for a great deal of the same things your ancestors used to use me for. But you only have those phones because of me.

It's not just that the physical materials used to make your phone were created when *my* stars died. All of the atoms in the phone—and in you, for that matter—were made in me. That Sagan fellow was correct; you are all made of star stuff. But the technology your phones rely on *also* exists because of me. Or rather, because of your scientists' fascination with me.

Every time you use your phone to find the nearest coffee shop—seriously, what makes you so tired that you need that much coffee? I make at least five new stars and move ten billion miles every year, but you don't see me chugging caffeine every morning—you interact with satellites. Your phone receives radio waves (which you can't see because your eyes are so tragically small) from multiple satellites at once and uses the slight differences in the signals' arrival times to pinpoint your location.

Are you following, human?

It doesn't really matter. The important thing is that without satellites, you wouldn't be able to navigate your tiny rock. You also wouldn't have high-speed internet, long-distance calls, or—to get back to your oh-so-important coffee—the option to pay for your morning cappuccino with your credit card. And the only reason you have satellites in the first place is that human scientists wanted to study *me*.

After thousands of years of tracking my movement, your ancestors started to understand how motion, gravity, and light waves work. They used that knowledge to launch machines out of your atmosphere, and now you can call your international friend while simultaneously buying things online with money that you've never actually touched.

Beyond this recent global positioning technology, your broadening understanding of space has introduced other life-altering innovations like digital cameras, wireless internet, and noninvasive security checks like X-ray machines. Even the procedures your doctors use to sterilize hospital rooms so your delicate human bodies can stay free from contamination were originally developed to protect telescopes while they did the vital work of observing me.[7]

You're welcome.

That's enough about you for now. It's time for more important things. It's time for you to learn something about me.

CHAPTER TWO

MY NAMES

I INTRODUCED MYSELF AS THE Milky Way because that's what most of you call me now, but that's not how you've always referred to me, and to be clear, it's certainly not what I would have called myself.

Humans have given me so many names over the years—Milky Way, Silver River, Way of Birds, the Deer Jump—and almost all of them can be traced back to myths from around your tiny little rock. The subject of the stories may have stayed the same, but the content varied based on the local customs and surroundings of whoever told it. Many human cultures saw me as spilled milk spread across the sky, but there were also some who thought of me as flowing water, scattered straw, or windswept embers.

After so many billions of years of destroying any new thing that came close to me, it felt nice to be called the Straw Thief's Way. Humans have bizarrely strong feelings about possession, so you probably wouldn't jump to be associated with a thief, but the early

Armenians saw this particular theft differently. They told stories of a legendary frigid winter when Vahagn, their god of fire, took pity on them and stole straw from the king of neighboring Assyria so they could keep themselves warm. You and I both know that straw isn't the most effective fuel for fire, but Vahagn, being born from a reed of burning straw, had a personal connection to it. As he fled from Assyria with his god-sized arms full of royal straw, Vahagn dropped a smattering of reeds across the sky, because of course that's where gods travel. Allegedly, I am that path of life-saving straw. It was such a touching story that I didn't even ridicule them for thinking their winter was cold, even though it was hundreds of degrees warmer than the rest of the universe.

On the other side of your equator, the Khoisan of southern Africa told the story of a young girl who lived under a pitch-black sky. One night, after dancing around a fire, she realized that she was hungry but didn't have enough light to find her way home for dinner. But the best characters in any human story are resourceful and innovative, so she flung the embers from her fire across the sky to light her way home. Yet another altruistic act on my part, despite not being entirely by design: providing enough light to see by when your sun isn't around. Although technically, since your sun is a part of me, I generously provide light for you all the time.

Some humans of northern Europe call me the Bird's Path or Way of the Birds after they noticed that their birds followed me when they migrated south every autumn. That's right, I don't just help humans; you're not special.[1] My splendor inspired those humans to tell stories about the bird queen Lindu, a white bird

with a human woman's head. In all my years of monitoring my planets, I've never seen such a creature, but I don't mind a few fits of human fancy. Lindu's job was to lead the migrating birds to safety, but she was distracted from her duties by a broken heart. Typical human nonsense, thinking that a little rejection is enough to keep someone from performing the most essential tasks. Anyway, according to the myth, Lindu was abandoned by her betrothed before their wedding and cried so much that her father, the god of the sky, took pity on her and summoned her home. As the winds carried her away, Lindu's tear-soaked veil turned into millions of stars marking her path.

These myths, the names and other words your ancestors used to describe me, were all reflections of what your ancestors knew about the world around them. That's what all your myths are: tools for understanding the natural world and communicating that knowledge to others. Okay, there were some that were purely entertaining, but most taught a lesson of some sort. Many of you don't realize this, but myths were some of your species' first attempt at scientific inquiry. Hundreds of years after humans told the story of Lindu and her birds, your scientists found empirical proof that some migratory birds *do* navigate by my light.

I have loved watching your myths bleed into philosophy and then evolve into scientific explanation. As you learned more about me, I really did feel like we were growing closer, but I'll say this: you would save so much time if you just paid attention to what your ancestors knew long ago.

Most modern astronomers dismiss old stories about me as nonsense, but they still turn to mythology when they need to name a new object. You can see that pretty much everywhere you look, from the godly names given to the other planets that orbit your sun to the constellations you've cobbled together from objectively disconnected stars. Regardless of what inspired them, the names of all space objects must be approved by one organization: the International Astronomical Union. This IAU has taken it upon itself to be the official keeper of names in space, even though they've certainly never consulted with any of us celestial figures about what names we might prefer.

Despite my myriad names, or perhaps because of them, the IAU has avoided giving me an official moniker. They simply refer to me as "the Galaxy" in formal documents.

It's better, though, that you can all call me whatever you want, that your cultural stories and the knowledge they hold haven't been stripped from you by some tiny organization. After all, it was those stories that brought my attention to you humans in the first place, and I'd be sorry to see them lost to your short collective memories.

So call me River of Heaven, the Road to Santiago, Winter Street, or by another name that feels right to you. Just make sure that when you do call me, you have something intelligent to say.

CHAPTER THREE

EARLY YEARS

A VERY WISE WOMAN AND one of my favorite human performers (a true star—which means a lot coming from me) sang that the very beginning is "a very good place to start."[1] Indeed, most of your human autobiographies start with the writer's birth and then proceed chronologically. That's because, for your kind, it's easy to know when the beginning is. And yet I've watched so many of you freeze in terror when your child asks the dreaded question: Where do babies come from?

The question isn't new. Your ancestors figured out with remarkable speed how to make miniature, imperfect copies of themselves, and that knowledge had to be passed on somehow. The *way* you answer the question, however, has changed over time. A common answer these days involves birds, bees, and sometimes a stork for some reason? The small ones never walk away from those conversations with an accurate grasp of how they were conceived, but they seem satisfied, nevertheless.

There are no birds, bees, or storks in space, either literally or figuratively. And I don't have parents to ask how I came to be. But I have my memories, even though they get fuzzy when I look back at my earliest millennia (don't judge me; I'd wager you don't remember everything about the day you were born, either), and I've seen other galaxies form over time. That might sound like a very intimate thing to observe, but as I said, I've been bored. Plus, we in space don't have the same hang-ups about privacy that you humans do. Perhaps it's because we don't have any of those fleshy appendages to be embarrassed by.

My point being that I know enough about how I and other galaxies formed to know that I don't have a birthday. There was no single moment that split the timeline of the universe into *before Milky Way* and *anno Milky Way*—that means "in the year(s) of the Milky Way" since I know your fickle world has moved on from that particular language—just like there won't be a single moment that transitions us to *after Milky Way*.

I formed slowly and in pieces, pulled together by my own growing gravitational force. I'm *still* growing thanks to that gravitational pull.

So instead of starting with *my* beginning, I'll take Ms. Andrews's melodic advice and start at the *very* beginning, at least as far as any of us are concerned: the moment your scientists have dubbed the Big Bang.

Don't concern yourself with thoughts of what came before the Big Bang. That kind of knowledge is not for the likes of you—or even me, though I am fabulously worthy on nearly all other counts—to understand. And trying to will just give you a headache.

No one knows for sure what triggered the Big Bang, not even the most erudite of galaxies and especially not your human scientists with their minuscule, squishy brains. But the most widely accepted estimate is that it occurred about 13.8 billion years ago, plus or minus about 40 million years. That might seem like too much uncertainty to a creature as short-lived as yourself, but it is insignificant compared to galactic timescales. Before the Big Bang, which is a difficult time for all of us to conceptualize, all the matter and energy we can see in the universe was contained in an infinitesimally small point.

Finally, something on a small enough scale for you to understand!...Maybe? The limits of your perception make no sense to me.

When the Big Bang happened, all of that matter and energy were thrust outward. Human scientists don't know why or how it happened, though some of them are convinced they're on the very brink of figuring it out. The early universe was such an active time that some of your physicists have written entire books about the first few minutes. If you ask me, they missed out on all the interesting bits—me—by stopping the story so early, but they made a choice and I must respect their focus.

In the first tiny fraction of a second, the universe rapidly inflated to 100,000,000,000,000,000,000,000,000 times its original size. That number is 100 septillion, or 10^{26}, or 1 followed by 26 zeros. Expanding so rapidly allowed the universe to cool off by a factor of 100,000. Now when I say things like "hot" and "cold," know that I don't feel temperature like you do. The temperature of a space is ultimately a measurement of how fast particles in that space are moving. If the particles are moving quickly, the space has a higher temperature. And temperature, density, and volume are all related to each other, so as the universe expanded, it also became less dense. The particles slowed down and everything became markedly cooler.

Within minutes, the universe had cooled from about 10^{32} degrees Kelvin at its hottest to only about a billion degrees. Oh right, you don't all use the same temperature scale, so I'll tell you that 100 degrees Celsius or 212 degrees Fahrenheit (whichever you prefer, though I'll never understand why you can't just pick one) is only 373 Kelvin. Just imagine how hot a billion Kelvin feels.

A billion degrees is a crucial benchmark because that's when the universe became cool enough for the protons and neutrons that had formed to combine in simple groups of what human scientists would one day call atomic nuclei. Your scientists term that whole process Big Bang nucleosynthesis. I just call it making the first elements, but I don't need to try to sound intelligent merely to impress.

The universe was still so hot that electrons were moving too rapidly to bind to those first atomic nuclei. You've certainly never had to think about something being so hot that atoms can't form. The hottest thing you ever interact with only gets hot enough to prepare your dinners, not rip apart fundamental particles. That's probably a good thing for your fragile bodies.

In the wake of that impressive initial expansion, which your astronomers have so creatively dubbed the epoch of inflation, it took the universe hundreds of thousands of years to cool down enough for electrons to join the nuclei and form neutral atoms.

Those first atoms were mostly what you call hydrogen (the easiest to make because they require only one proton and one electron), some helium, and the tiniest amount of lithium.

I wasn't alive to witness this myself, but it was around that time, three hundred ninety thousand years after the Big Bang, that the universe became transparent. Before that, photons (particles of light) kept bouncing off the throngs of free electrons that hadn't yet bound to the atomic nuclei, and the universe looked opaque. I know this because looking out into space is like looking backward in time. Light, as fast as it is, can travel at only a finite speed. It takes time to travel the vast distances of space. And when I look back far enough, through space and therefore also time, I get to a point where I can't see anything. The universe looks dark because all the light was trapped.

But *seeing* isn't the only way to gather information. You humans have always relied too heavily on your sight, especially when there's so much to *feel* in space. Take the wealth of heat and energy present at the very beginning of the universe, for example. It didn't disappear, it just dispersed. And we can still detect the heat signature today, all around us in the universe. Your astronomers call it the cosmic microwave background, or CMB. If you're reading this actively—by which I mean you're thinking about what you read instead of just letting the words pass in one ear and out the other, figuratively speaking unless this is one of those audiobooks—you might be confused by the name of that radiation because *heat* is usually observed in the *infrared* part of the electromagnetic spectrum, not the *microwave*.

Do you know what that spectrum is? Oh, how your scientists

have failed you. The electromagnetic spectrum is the range of possible wavelengths for light. Radio waves have very long wavelengths and low energies, while gamma rays have short wavelengths and high energies. Your kind can see only in a very narrow band of this spectrum in between those extremes. What a waste.

Back to my point about the CMB: heat usually shows up as infrared light. But it's called the cosmic *microwave* background because the universe has expanded since the Big Bang, and the wavelength of that early light has expanded as well, pushing it out of the infrared region of the electromagnetic spectrum and into the microwave.

The CMB shows tiny temperature fluctuations that point out which spots were ever so slightly warmer, and therefore denser, than their surroundings. That pattern tells us—your top scientists and literally any galaxy with half a cosmic brain—what the temperature of the early universe was and how matter was distributed when the universe was still opaque.

Sadly, your ignorance compels me to explain so much to you that I'm still not at the part about me yet. But we're getting close! It took another three hundred million years after the first atoms formed to create the first stars. The universe's entire store of hydrogen and helium—and, yes, let's include lithium, too, just for accuracy's sake—existed as clouds of gas. Something disturbed the delicate equilibrium of those first clouds. Maybe some cosmic wind passed through them or maybe the gas was randomly distributed in such a way that led to one part of the cloud being denser than the rest. In fact, there are some human astronomers who study the tiny fluctuations in the CMB and run computer models to see what kind of large-scale structure is created as a consequence of the universe's initial pattern of under- and overdensities.[2]

Whatever caused it, as soon as there was some unevenness in the way matter was spread through the cloud, gravity was able to take over. It's different if you're talking about the infinitesimally small (atomic nuclei) or the inconceivably gargantuan (the expanding universe), but on most size scales, gravity controls *everything*. That small overdense region attracted more and more material until eventually it collapsed in on itself under the force of its own weight, getting hotter and denser until it created the very first stars.

The creation of those first stars sent powerful shock waves through the rest of the cloud, and the stars themselves disturbed their surroundings by heating them up, ionizing the gas with their radiation, and giving off winds of charged particles. Those disturbances set off a cascade of star formation like you humans knocking down a line of dominoes. That same process happened in gas clouds nearby, and over time, all those separate bundles of gas and stars and dark matter were drawn together by gravity. When they met, they merged, sharing their stars and creating new ones as their gases mixed together.

Those early stars composed of primordial hydrogen and helium burned hard and fast, exhausting their hydrogen fuel in only tens of millions of years. Indeed, human scientists haven't been able to find any stars from this first generation, which they confusingly call Population III stars. (They call youngest stars Pop I, even though they're from a later generation.) Every star they observe has at least some contamination by metals, though it is possible that some of the first stars survived and have picked up metals in their outer atmospheres as they travel through clouds of enriched gas.

That's what your astronomers call elements heavier than helium, by the way: metals. Most of your other scientists seem to have a very different idea of what a metal is, but I'm not here to get in the middle of such silly linguistic squabbles.

The first generation of stars produced some heavier elements[3] in their cores—what you call beryllium, carbon, nitrogen... all the way up to iron—and when they died, they released those elements out into space to be used in the next generation of stars that were just a bit more metal-rich than their predecessors.

Let's pause for a moment, human, because I don't want you to get the harebrained idea that my stars form in neatly organized and scheduled batches. The truth is that I'm forming stars all the time and stars are, sadly, dying all the time. I'll go into more detail about that later, but for now, take the phrase "generation of stars" literally. New humans die and are born every year, but you still say that a new generation comes about roughly every twenty-five years based on average characteristics of people born during that time. My stars are just the same.

Over hundreds of millions of years, the cycle of cloud collapse, metal production, and gravitational attraction produced the earliest examples of galaxies. They had all the stuff that a galaxy is supposed to have—stars, gas, dust, and dark matter (my stunning good looks are just a bonus, not a requirement)—but they were smaller than I am now. And we grew by eating each other.

First, don't get hung up on the idea of galaxies eating each other. It's just what we do—no more disturbing than consuming pineapple on pizza. And second, I'm starting to say "we" because at this point, several hundred million years after the Big Bang, most of my parts were already created and it was only a matter of time before gravity brought them all together. Before it made *us* into *me*. So, congratulations! We've finally arrived at my beginning.

I know it's likely a lot for you to take in, and your brain is fully formed! So, if a child asks how galaxies are born, you can tell them that when a gas cloud loves itself very much, it hugs itself extra tight, and after a few hundred million years, a baby galaxy is born. Leave the storks out of it, please.

Back then, all of the little galaxy bundles—protogalaxies, as your astronomers might call them—were young, hot, and packed into a much smaller space than we occupy now. We had contests to see who could eat the most gas and who could form stars the fastest—the usual youthful debauchery—in part because it was fun to compete, but mostly because we knew that the biggest galaxies were the ones who would survive. Those first few hundred million years were one big, wild, high-stakes party, not unlike your Burning Man.

The party was only possible because the universe was so much hotter and denser then than it is now. The average temperature back then was about 50 Kelvin, which is cool even by your standards, but the fast-moving particles of the early universe made it easier to form, swap, and merge material with other galaxies. As the universe continued to expand, propelled by a mysterious force that your scientists don't understand at all, it cooled down. (Though your astronomers have recently discovered that the gas in the universe has been heating up over the last ten billion years as gravity clumps it together.[4])

Nowadays, the universe is a cold and still place, almost as cold as anything can be. Looking again at the CMB, we can tell that the average temperature of the universe is just 2.7 Kelvin—that's twenty times colder than it was when I was born!

By the way, I'm specifying the *average* temperature here because there are parts of the universe now that are much hotter and cooler than 2.7 K. An average star like your sun, for example, has a temperature of 5,800 K, while your human body maintains a constant temperature of only 310 K. It's only on sufficiently large scales—even larger than me—that the temperature of the universe is that low. And as far as I know, nothing in the universe has a temperature of 0 K, the temperature associated with no particle motion at all, or "absolute zero," as you call it.

Even if the universe never actually reaches absolute zero, it was able to cool down by nonillions of degrees in just three hundred million years. That's a difference of more than 10^{30} degrees in less time than it took bacterial life to form on your planet.

You humans sometimes throw these numbers around—three hundred million, billion, nonillion—but I don't think you could possibly have a good grasp of what they mean. Most of these numbers are paltry to me, but you so rarely interact with values that large in your short lives. In fact, some human languages don't even have words to distinguish between them, and would merely call them all "large numbers."[5] But to show you how different those numbers are, let me shrink it down to something you're more familiar with. Instead of years, imagine I'm talking about *seconds* after the Big Bang.

Three hundred thousand seconds is three and a half Earth days, and that's when the universe first started making atoms. It took another three hundred million seconds, or another TEN YEARS for stars to form. And now we're here, nearly fourteen billion seconds after the Big Bang, which is almost four hundred fifty Earth years.

You can also imagine that the universe cooling down so considerably after the Big Bang to the point where stars could form is like your sun turning into a ball of ice in just three days. The universe was able to cool down so quickly because it was expanding at such a lightning rate, keeping the same amount of matter and energy in a rapidly growing space. That expansion, which continues to this day, is the reason there are so few galaxies like me left in my neighborhood. There are still traces of light that I can see from most of the galaxies who abandoned me. Many of them have grown into full galaxies and built their own little galactic communities just like the neighborhood where I (and you) live, but they're moving farther away all the time. One day, I'll look out and they'll all be gone. Don't worry, they won't be dead. At least, most of them won't be. They'll just be so far away that their light can't reach us anymore. But that shouldn't happen for another one hundred billion years, so there's no reason for either of us to dwell on it now.

Most of the galaxies who are still around are dwarf galaxies, which I suppose I must describe because there's some potential for confusion about the threshold between a dwarf galaxy and a big galaxy like me. Several years ago, there was a large controversy when one of your beloved planets in your solar system was demoted to a dwarf planet. Astronomers claimed it was because the icy chunk of rock wasn't massive enough to have cleared out the debris in its orbital path. I have no quarrel with that; what you call a planet is up to you. What do I care? I have hundreds of billions of planets.

I wonder, though: Would you have the same kind of outraged reaction if I were demoted to a dwarf galaxy?

It's a completely hypothetical question because I crossed the threshold from dwarf to galaxy long ago. I can't know exactly when it happened because human astronomers don't agree on what the cutoff between dwarf and not-dwarf should be. Some of them use a mass limit, while others use size, brightness, shape...Nearly every astronomer who cares about dwarf galaxies has their own idea of what defines them. Of course, that's inherently frustrating, but in most cases, it's fairly obvious which category a given galaxy falls into. Dwarfs are small galaxies with only a few hundred million stars. The biggest ones have maybe a few billion.

Their small stature is, naturally, a consequence of where, when, and how they were born. Aren't we all?

Some dwarfs form through gravitational interactions, or tidal forces, between bigger galaxies. When galaxies have a sufficiently violent interaction—such as when they try to eat each other but the losing one puts up a fight—they might fling some material away from the fray. Actually, the same thing happens when galaxies have intensely, um...intimate interactions. So, perhaps some dwarf galaxies really do have something like parents.

Other dwarf galaxies formed around the same time as my inception. These so-called primordial dwarfs are poor in metals and slow with their star formation, but it's not always their fault.

Many of these primordial dwarfs had their gas stripped and star formation quenched by their central black holes. It's a shame, really. But there are some primordial dwarfs who remained small because they just didn't work hard enough or eat enough material when they were young.

I'm not implying that I'm better than dwarf galaxies just because of my size. There are some galaxies right on the cusp between dwarf and not, like the Large Magellanic Cloud, whom I know better as Larry. The two of us have our disagreements, and I am the objectively better galaxy, but it's not because of the difference in our sizes.

One fatal flaw for dwarf galaxies is that, because they have less mass than a full-grown galaxy, they're easier to rip apart with gravitational forces. I see this happen all the time to smaller galaxies around me. I've even ripped apart several of them myself. If I didn't, I would die. And they know it's for the best, too, because it gives their stars a more stable home.

But not all dwarfs get ripped apart. If they did, we wouldn't see any around today. Even I will one day be ripped apart by the gravity of a larger structure, so I'm not necessarily superior to them. I'm still glad I'm not a dwarf, though.

Alas, it seems I've gotten a little carried away, but all you need to know is that I grew big enough to be unequivocally *not* a dwarf galaxy because I worked hard to bulk up, and gravity is king. Or queen. Who knows? Galaxies don't care about gender like you do. None of those fleshy bits, remember?

To bring it all back to where we started, I've been around for about 13.5 billion years. My early gigaanna (that means multiple periods of a billion years; learn some Latin) were spent gorging myself on gas and ripping apart smaller galaxies who may or may not have been dwarfs. (Honestly, once you go far enough back in time, human classification schemes become even less useful than they are now.) That's billions of years of trying to strike a balance between creation (stars, planets, black holes) and destruction (supernovae, gamma-ray bursts, and tidal ripping). Matter in its most fundamental form can never be destroyed, but lives certainly can, and I've unfortunately ended enough cosmic lives to tip the scales in the wrong direction.

I do know I felt a lot better after you showed up. Not you specifically, obviously. You as an individual aren't that significant, but you humans as a group are.

Take, for example, the strides you've made in your understanding of the universe. You certainly have a long way to go, but there was a time when your kind believed that the night sky was a rock with holes pricked in it by all-powerful creatures.[6] And just a few thousand years later, you took an actual *picture* of a black hole in another galaxy! All without ever traveling from your tiny little position in space and time.

I can't stress enough how surprising that is. You have a small animal on your planet—a mayfly, I think you call it—that lives for one Earth day if it's lucky.[7] A mayfly can live its entire life in one room of one of your homes. Isn't that sad? Don't you ever wonder why the mayfly even bothers to do anything at all? Because that's how I feel about you. Your lives are so short and confined that many of you can't even fathom the extent of my existence, but *you keep trying*. If I were in your shoes, even with my impressive patience, I would have given up long ago.

As you developed methods and tools to study me rigorously and uncover explanations for what your ancestors saw with their naked eyes for millennia, you kept running into the same question: How old is it? (Odd, since this seems to be a rude question on your planet.) You were interested in learning about the nature of things—how they formed, evolved, and died—but to know how something has changed over time, you have to know how much time it has had to change.

There are some of you who, after reading a book written more than a thousand years ago in a language you can't understand, claim that I'm not even ten thousand years old. There are also many of you who don't really care how old I am because of a belief that it doesn't affect your lives. I may be biased, but I'd say my age affects all of you because you live in me. If I were much younger, for example, there might not be enough carbon and calcium in my gas clouds to make humans in the first place.

But there's a small group of you who have devoted their lives to figuring out how old I and my various parts are, and how I've changed over time. Those curious humans call themselves galactic archaeologists.

You were lucky enough to find out my age just by reading this book, but galactic archaeologists had to learn my age on their own. And it was certainly diverting to watch them come up with the answer themselves.

Much like an Earth archaeologist who determines the age of an ancient (to you) civilization by figuring out how old a clay pot is, galactic archaeologists realized they could determine my age by measuring the ages of my oldest stars. They were so creative—

a trait shared by most of your kind—and came up with several methods to do just that. And, of course, I have a few favorites.

One of them depends on models, meaning the astronomers think they know how stars work so they can make good guesses about how a star is going to change over time. With this method, you can measure how hot and bright a particular star is and use those values to match the star to one of the models. Human astronomers call this isochrones fitting. The word "isochrone" comes from two words in an old human language meaning "same" and "time," which is appropriate, because astronomers use them to identify stars that were born at the same time.

I was so tickled when the astronomers came up with this method because *they knew it was unreliable*, especially for stars less massive than your sun. The models depend on knowing values like the mass and temperature of a star, which have their own uncertainties that get carried through to the age calculation. When astronomers compared their isochrone ages to those found with other methods, they realized that isochrones can be wrong by as much as a factor of two, but they're usually just wrong by about 25 percent.

It might seem cruel that I was so amused by your wrongness, but it's actually the opposite. For a long time, this was the best way you had to find my stars' ages. And, knowing that your best wasn't good enough, you kept working to come up with better methods. Your kind is tireless, working hard to achieve what you can with your short lives and limited resources.

A slightly more precise method of aging stars depends on values that you can directly observe and measure. Some of your astronomers figured out that my stars all rotate about some axis just like your Earth does, and that they start to spin more slowly as they get older. Unlike your planet, which is slowing down because of gravitational interactions with your moon,[8] stars slow down because they give off magnetic winds that drag at them as they spin. Some stars—I'm made of different types of stars just like you're made of different types of cells—are born spinning faster than others, but they also have stronger winds that slow them down faster, so they eventually catch up with their slower-spinning peers. The ones that "spin down" like this are all less massive than your sun.

Human astronomers have observed, modeled, and simulated stars slowing down, and now they can calculate a low-mass star's age based just on its rotation speed with a technique they call gyrochronology.

Beyond fitting isochrone models and ogling my stars as they dance, your scientists have tried to age stars by tracking changes in their orbits around my body, timing how fast they pulse (get bigger and smaller, like your chest when you breathe), and measuring how much lithium they hold. Most of these methods[9] give highly uncertain answers for individual stars, but when you apply them to entire cohorts of stars, the answers become more precise. (Don't ask why; I'm not here to teach you *statistics*.) Your astronomers call these groups of stars born from the same gas cloud "open clusters" because they're usually sparsely populated enough to dissolve—or break *open*—over two hundred million years or so.

On the off chance you've ever thought about the ages of stars before, it was probably because you care about how they affect humanity's relentless search for life on other planets. I swear, finding aliens is like your species' favorite hobby. A planet's evolution is tied to its star's, so the key to finding life is knowing whether a system has existed for the right amount of time to host it.[10] Your astronomers even have a saying: "Know thy star, know thy planet." If you figure out how old, hot, and metallic a star is, you can infer a lot about its planet(s). Without giving too many of my juiciest secrets away, I can tell you that your star was born at just the right moment to have the right ingredients for life and enough time to build it...give or take a few hundred million years.

You can trace your timely existence all the way back to a random fluctuation in a primordial gas cloud thirteen billion years ago. Without that tiny disturbance, the first stars wouldn't have formed, I wouldn't have been born, and my stars wouldn't have produced enough carbon in their cores to make you. Hopefully, learning about these unfathomably long timescales of the universe will help you finally understand how ephemeral your existence is in the grand scheme of things. But also, every single proton, neutron, and electron in your body was made in the first three minutes after the Big Bang. That must blow your little human mind!

Good. Now let's do the same thing again, but with size instead of time.

CHAPTER FOUR

CREATION

BEFORE I GET TO THAT, do you understand how lucky you are to be learning this kind of vital information directly from me, an *actual* galaxy? You'd probably be just as nonplussed if it were that almost-dwarf Larry writing this, though I guarantee you wouldn't find Larry's explanations nearly as entertaining. My telling you this story—my story—is a *gift*. It's like if you learned about...oh, what's something you humans admire? It's like Beyoncé taking time out of her "busy" schedule to personally give you singing lessons. Even that falls short, though—she's not supervising a hundred billion stars.

Your ancestors didn't have this book, or the fancy machinery your scientists use, or the thousands of years' worth of accumulated knowledge that you benefit from. They didn't know anything about the truth of the Big Bang. Instead, they had gods: powerful, immortal, otherworldly beings who created and maintained the ever-changing universe. Your ancestors drew the best conclusions they could from the information available to them through their weak human senses, just like you do. Or at least just like you should.

That hard work of trying to make sense of the world around them gave them a healthy respect for yours truly. And while I'm neither a god nor a believer in any, I still appreciate a good story, especially one that has just a kernel of truth in it—even, selflessly enough, if it doesn't include me. But let's be honest, the stories with me are always better than the ones without. While I could tell you about the most popular or widely believed creation myths, your lives are exceedingly short, so I'll skip to the ones I favor.

I mentioned an ever-changing universe. Hopefully you comprehend by now, owing to science and the marvels of modern publishing, that the universe is changing, morphing, expanding. If left to learn from first principles, you'd think the universe was fixed and constant, because that's how it looks from your limited human perspective. And yet, somehow, some of the creation stories your ancestors told describe a universe that's in constant flux, operating on an infinite cycle of birth and destruction. Some of your modern astronomers tell a similar story, but they tell it with math and computer code instead of words.

One of these cyclical cosmogonies came from the people of your Indus River Valley more than four thousand years ago. They practiced a religion called Hinduism, the oldest of your planet's most popular current doctrines. Hindus believe that the god Brahma created the cosmos himself—words like "universe," "world," and "cosmos" were more or less interchangeable before they adopted their modern scientific definitions—and that ours is not the first one he created.

Brahma is far from the only god in the Hindu religion. In fact, the idea that there's a single true god is relatively new. There is also Vishnu the preserver, who maintains the balance of the cosmos. It's no wonder that Vishnu was often associated with the sun, as both were understood to sustain life on Earth. To round out the cycle, there's Shiva, who destroys the universe so that it can be rebuilt. But until that time comes, Shiva is said to destroy the imperfections of your world, and so he is regarded as both good and evil. Together, the three gods, this *triumvirate*, work together to keep the universe moving through its cycle, each doing their part when the time comes, until the end of eternity. Or, if I know anything about immortal beings, until they get bored of doing the same thing over and over. But maybe I'm projecting.

Three thousand years later and 4,500 miles to the north, Norse tribes were telling their own cosmogonies that were somewhat rooted in truth. The stories were passed orally through countless generations, your imperfect human memories and pesky personal preferences introducing slight variations each time, until they were written down in your thirteenth century. Christianity was well established in the north lands by then, and it's hard for even me to say how much the *Prose* and *Poetic Edda*s differed from the pagan stories that early Vikings shared around their fires. Honestly, I wasn't paying much attention. Humanity's Middle Ages were boring and I had other stuff to do.

The Eddas describe a great abyss that stretched between the first two worlds: Muspelheim, the world of fire, and Niflheim, the world of ice. Frost and flame met in the middle, and a giant god was born of the melted ice. His name was Ymir, and he was later slaughtered by some of the creatures who sprang from his body, and his parts were used to construct the other worlds in the Norse universe. There are nine of them in total, including separate homes for humans and their gods. These worlds were supposed to have rested among the roots and branches of Yggdrasil, the great world tree.

I'll soon share more details about my body and the shape of the actual universe, but suffice it to say that on no scale is space shaped like a tree. Well, it might kind of look like tree roots if you zoom out far enough.

Still, the Norse story has that improbable kernel of truth that I love to see. Life emerged in the middle of the abyss, between the worlds of ice and fire where the temperature was just right. Right for what, you ask? Liquid water, of course. You know, that sloshy stuff that you're all full of and *so* dependent on. The Nordic people, having lived on a literal land of fire and ice (volcanoes and glaciers), would have witnessed how life can flourish where the two meet. Water, like the sun, nourishes your frail little bodies, so it, too, often makes its way into your most sacred stories.

So many of the creation stories your ancestors told started not with chaos or nothing, but with a deep primordial ocean. My favorites of these involve a divine creature diving to the bottom of this ocean to gather bits of mud that then get used to build the land. The diver often takes the shape of an animal of some kind, a wonderfully whimsical picture, and many of the stories include multiple failed attempts before the mud at the bottom of the ocean is successfully retrieved.

This kind of account, sometimes collectively called Earth Diver myths, is common among the Indigenous people of North

America. But similar stories can also be found in modern Turkey, northern Europe, and eastern Russia. Some of the humans who spend their short lives tracking the evolution of your ancestors' stories—you call them folklorists or anthropologists—believe that the Earth Diver myths share a common narrative ancestor from eastern Asia that spread as the people migrated.

Now, clearly the Earth Diver story as a creation myth is focused on the creation of Earth's land, centering your meaningless little rock. You might think that would put me off it, but you'd certainly be wrong. For all intents and purposes, Earth *was* your ancestors' universe. Life on Earth *did* originate in the water. And humanity *is* the latest attempt at life after so many catastrophic failures. More species have gone extinct on your planet than there are living now.[1] (RIP to the trilobites. I had big hopes for them.) So, the Earth Diver stories get a lot right.

I never expected your ancestors to know everything about me. They obviously appreciated my presence, so I was content to listen to their stories and watch as they marched steadily towards science without knowing what they would find. It was entertaining. And maybe even a little bit inspiring.

But *your* ignorance of the vast universe around you is neither. You have the tools and the experts and the knowledge all available to you, but you haven't used them. Hence, my decision to finally intervene.

Now, as you read the rest of my story—which, again, is a privilege—remember that you're no smarter than your ancestors who believed the sky was made from a dead giant's skull. You were just lucky to be born later.

CHAPTER FIVE

HOMETOWN

How do you feel about your hometown? I've noticed that some of you become attached to your football teams (though "football" means different things to certain parts of the world because you're a ridiculous species) and your regional foods. I have often found myself wondering: What is a cheese-steak and what does it have to do with horses? The rest of you seem to work exceedingly hard to get as far away from your hometown as possible. It's never made much sense to me because, from where I'm looking, you all come from the same place, but I guess an ocean would look big and obstructive to something puny like you.

So, allow me to introduce you to my hometown, which, by extension, is also yours. You're welcome to feel proud of it. In fact, I encourage you to, but you should certainly learn more about it first. As a species, you humans make too many decisions without having all the facts.

There are, of course, other galaxies besides me out in the universe, all of them less spectacular than me, with one remarkable exception. Most of those galaxies are tens of millions of light-years away and getting farther every second. But some galaxies—your astronomers have discovered about fifty so far—sit right beyond our galactic backyard, so to speak. As with any neighborhood, its quality depends on the people who live here, and we have a rather mixed bag.

They—we—are all gravitationally bound to each other. Only the most extreme case of universal expansion could drive us apart, and even that would take a couple dozen billion years. We are both literally and figuratively stuck with each other. Your human astronomers call our little gaggle of galaxies the Local Group.

The Local Group is about ten million light-years wide, surrounding its two most influential members: the galaxy you call Andromeda and, of course, me. Every other galaxy in the group is smaller than me, except maybe Andromeda, and most of them are less than 1 percent of my size. Those smallest ones don't do much besides orbit larger galaxies with more gravity and responsibilities. They're those neighbors who can't be bothered to participate in anything—they don't decorate for holidays or bring anything to the cookout or volunteer for the neighborhood watch or whatever else you humans do in your little communities—but they still soak up some of the resources. Except in our neighborhood, the only resource we really care about is intergalactic gas.

I try not to think about these annoying hoverers much. Focus on them for too long and their persistent buzzing around my halo gets uncomfortable. You corporeal beings might call the sensation "itchy." Your astronomers have observed the way my disk squirms and wiggles in response.[1]

One particularly egregious pest is the dwarf galaxy your astronomers call Sagittarius, which ventured too close a few hundred million years ago, irking me until I started to tear it apart with my gravity. Now its stars are scattered around my body in the so-called Sagittarius stream,[2] and I have its gas to snack on for eons to come. Smaller galaxies take longer to consume because their low masses make it difficult to get a good gravitational hold on them, but their defeat is always inevitable. Why bother making nice or even friends with a galaxy when I know I'll just end up eating it in a billion years. Have you ever bared your soul to a bowl of hard candies?

But there are a few neighborhood galaxies who, over time, became what I guess you might call my friends. The biggest and brightest and most important of them is Andromeda, but I won't give away too much of that story just yet. That leaves Larry, Sammy, and Trin, or as you probably know them, the Large Magellanic Cloud, the Small Magellanic Cloud, and Triangulum.

"Friends" is possibly too strong a word, and not always accurate. Do you have a word for a being whose presence you tolerate because its absence would lead to mind-shattering loneliness and despair?[3]

You humans can actually spot all three of them with your weak naked eyes. Or at least your ancestors could see them before your species belched all kinds of pollution into Earth's atmosphere.

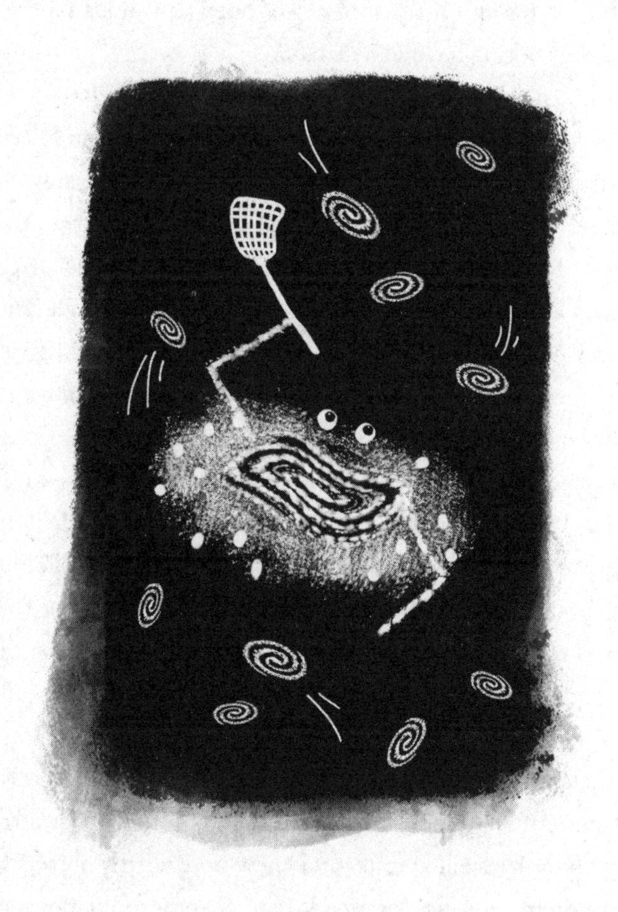

Now you'd have to make a concerted effort to travel to somewhere dark enough to catch a glimpse. Don't feel bad if you've never made the trek. Only one of them even mildly deserves your attention, anyway. Of the other two, one is a jealous failure and the other is boring beyond belief.

Let's start with the worst of them, shall we?

Like me, Triangulum[4] has multiple names: Messier 33 (which your astronomers often shorten to M33), NGC 958, sometimes Pinwheel Galaxy, and my personal favorite, Trin the Underachiever. That last one isn't recognized by your precious IAU, but it should be because it's true! Trin is a little more than half my size and contains maybe a tenth as many stars, making it the third largest galaxy in the Local Group. This has always made Trin bitter.

I'm sure Trin would also be bothered—but certainly wouldn't admit it—if I revealed that ancient humans didn't include it in any of their stories. At least, none of the stories that were told loud or often enough for me to hear them, and they don't appear to have survived to be heard by you. Trin is simply too dim to be of any widespread use.

The first human who even bothered to write about Trin in a way that was preserved was an Italian astronomer in the 1600s. Giovanni Battista Hodierna was an accomplished court astronomer who took excellent notes and knew noteworthy objects when he saw them, and he described Trin as some unnamed nebula "near the Triangle." Any fuzzy patch in the sky was called a nebula back then, and the Triangle in question was the constellation now known as Triangulum. A little over a century later, Trin was added as the thirty-third item in the Messier catalog, a list of objects visible from your northern hemisphere by French astronomer Charles Messier. Hence the nickname M33. But do you want to know the best part? Back in your eighteenth century, Messier and his contemporaries were mostly interested in identifying comets, and Messier kept a list of all the frustrating objects that got in his way. And that's the list Trin's on! Ha! What a loser.

Trin sits on the other end of the Local Group, nearly three million light-years away from us, which is fabulous for me, but unfortunate for Andromeda, who has the—and I say this dripping with sarcasm, since I have no eyes to roll for you—*pleasure* of keeping Trin in orbit.

Trin has always sucked up to Andromeda, offering up streams of hydrogen and stars and following Andromeda like one of your lovesick puppy dogs. Unfortunately, Trin's sad little courtship will bear fruit over time, because if they keep on their current path, the two galaxies will probably merge in about two billion years. It's hard to be certain, though. The only thing I know for sure is that Andromeda would tear that sorry excuse for a galaxy to shreds and then remember that a much more appropriate companion is sitting patiently on the other side of the Local Group.

Until then, Triangulum will keep lobbing pathetic attempts at insults towards me:

"Oh, is that all the stars you can make in a year?"

"I have the brightest X-ray source in the entire Local Group."

"Your rotation curve is looking pretty steep."

Frankly, I couldn't give a fig about what Trin says. Being the third best—and by such a large margin—can't be easy. And

I'm so magnanimous that sometimes I even feel bad for the poor spiral.

But not so bad as to keep talking about it, so let's move on.

Have you ever met someone who just. Can't. Make. A decision? I suppose for you that would be someone who can't decide where to eat for dinner or whether to take some (probably average) job. Well, in our neighborhood, that's Larry, who can't choose between being a galaxy and being a dwarf.

Larry is the fourth largest galaxy in the Local Group and has neither delusions about climbing up that chain nor hard feelings about not being in the competition. Whereas Trin's petty personality comes from being just out of reach of greatness, Larry never had a chance to be anything special and therefore never established much of an identity at all.

But that doesn't mean Larry isn't a little impressive. No, a 14,000 light-year diameter and mass of more than ten billion of your suns is certainly remarkable, especially for a dwarf galaxy. Larry's just... wishy-washy. It's boring and it's just not the type of energy I like to have around. I know you can't really tell from the ground, so you'll just have to take my word for it as a galaxy who's watched Larry absentmindedly push dust and gas around for billions of years. Only pushing, because Cosmos forbid any stars actually get made!

Sporting only one spiral arm, Larry couldn't even fully commit to being a spiral galaxy. One arm! Your astronomers named a whole category of galaxies after Larry in what I can only assume was an act of pity. So now there's a whole class of "Magellanic spirals" with just a single spiral arm.[5] It's not even a *cool* class of galaxies because they are far too common to form any kind of exclusive club.

And Larry's always just *so close*. Fifty kiloparsecs; 163,000 light-years; 960 quadrillion miles—however you want to put it, it's not far enough away. Andromeda couldn't even squeeze between us. And that, my tiny little inhabitants, is just unacceptable.

If you ask me—and I know you didn't; that's why I'm doing this whole proverbial song and dance in the first place—the most interesting thing about Larry is the totally unexpected and undeserved bond it shares with Sammy. And I do mean literal bond. The two dwarf galaxies are joined by a stream of stars and gas stretching between them, the result of countless gravitational interactions—if I had eyebrows, I would be waggling them salaciously—over the last few billion years. And as surprising as it is to me, the union has been good for them. In the last two billion years, their star formation rates have increased dramatically, though they're both still underperforming for their masses. They seem happy, and I'm happy for them. Their lives are short enough that they should experience whatever joy while they still can.

My acquaintances they may be—I'd even consider Sammy the closest thing I have to a true friend—but none of us can fight our natures for very long. In a few billion years, my gravitational pull will lure them in, and I'll devour them both.

Don't pity them though. We've always known the time would come, and their stars will survive, even if they do get somewhat scattered throughout my body.

But I've gotten ahead of myself. I haven't even properly introduced you to Sammy yet.

You might know Sammy as the Small Magellanic Cloud, and perhaps you've even seen the blurry smudge in your night sky if you live near your planet's southern hemisphere. Sitting just two hundred thousand light-years away and weighing in at only seven billion times the mass of your sun, Sammy is one of my closest and, pathetically, largest neighbors after Larry.

Sammy is what your astronomers call an irregular dwarf galaxy. "Irregular," as far as I can tell, is a human astronomer's way of saying "misshapen blob" without sounding rude or overly colloquial. So couth, your astronomers. It basically means that Sammy doesn't have a beautiful spiral shape like I do, but not every galaxy can have my model morphology.

I should probably confess that I'm responsible for that irregularity. Sammy used to be a small version of a spiral galaxy with a strong bar connecting the arms in the middle—you can still see remnants of a bar if you look with one of your telescopes—but I got a bit miffed one day and tried to use my gravitational pull to rip Sammy apart. I swear it wasn't jealousy! I hadn't eaten anything in a few millennia. Even galaxies get hangry.

Both Larry and Sammy are so easy to see with your weak eyes—often together—that ancient humans knew and told stories about them long before anyone cared to write them down. Polynesian seafarers knew how to navigate by Larry and Sammy, the Maori in your modern New Zealand marked the clouds' return in their sky to predict weather, and some groups native to Australia looked to them as the resting place for the spirits of their loved ones. The knowledge of how the two dwarf galaxies could be used was passed from generation to generation by word of mouth, wrapped in stories to make them easier to remember. Most of those stories didn't involve me, though, except for one, also told in Australia. In it, Larry and Sammy were an old married couple known as Jukara. They were too feeble to find their own food, so they had to rely on the kindness of star people to bring them fish from the sky river. That sky river was, of course, me. Those ancient humans had no way of knowing that the exchange of sustenance between the three of us would one day go in the opposite direction.

If you've never heard these stories, you're probably from the northern half of your little blue dot, where Larry and Sammy are not as readily visible. The Greek and Norse myths that most of you seem familiar with wouldn't have mentioned them.

Once written language was developed, any human astronomer worth their salt—and salt was mighty valuable for much of human history, you know—wrote about Sammy and Larry, though they didn't use those names. It wasn't until what you call the sixteenth century—because apparently you ignore the forty-five million centuries that came before it—that humans started calling them the Small and Large Magellanic Clouds after some blowhard named Magellan saw them while he sailed around the globe.

For better or worse, those are the companions the universe has provided. Are you still with me, human? Good, because we have a lot more ground to cover.

I've been throwing out distances like two hundred thousand and ten million light-years as if you could possibly comprehend them. That's my bad—I know your planet is so small that you can't envision these magnitudes with your little brains. I don't even know whether your astronomers can truly fathom these vast distances, but at least they found ways to measure them.

From your limited perspective, the night sky looks two-dimensional. Indeed, some of your ancient ancestors believed that the sky was a blanket placed around the Earth but, like, a magical blanket with moving images on it. Your astronomers were clever enough to find a way to add that third and rather important dimension: distance. They call the sequence of methods they developed to measure farther and farther afield the "distance ladder."

The first rung on this ladder works only for nearby objects. Your astronomers call it the parallax method, and even under the best circumstances—looking at the brightest targets with the best telescopes—it can only reliably measure distances out to maybe ten thousand light-years. That's not even far enough to reach my closest galactic neighbor.

Parallax works by measuring how much an object's apparent position—meaning the position it appears to have on the sky—shifts as the observer moves. You've likely done this yourself on small scales, though you were blissfully ignorant of it at the time. If you hold your thumb out at arm's length and close first one eye, then switch to the other, your thumb appears to move, yes? That's parallax in action. The more distant the object, the less it appears to move. That's why this method of measuring distances gets less reliable for more distant targets.

"But great merciful Milky Way," you ask, "how did astronomers figure this out if we humans are stuck on our insignificant little planet?" At least, that's how I assume you would address me.

And the answer, human, is quite simple. You may be stuck on

your planet, but your planet moves around your sun. Astronomers measure the apparent shift of distant objects as your planet moves from one side of your sun to the other.

The parallax method even led your astronomers to create a new unit of measurement. They do this often, actually, but I think this one is worth the time and effort it will take to explain the concept to you.

The new unit is called a parsec. It's the distance from your sun to an object that has a ***par***allax angle of one arc***sec***ond. I have to explain arcseconds, too, don't I? Fine. An arcsecond is a unit used to measure very small angles. You've heard of degrees? Not the temperature kind, but the type that has to do with shapes. An arcsecond is one-sixtieth of an arcminute, which is one-sixtieth of a degree. You could call a degree an archour if you wanted, but only people who have read this book would know what you were talking about.

Wow, this explanation is turning out to be far more obtuse than I thought it would be. (That's a little angle humor for you. How very a-cute of me.)

The parsec is more commonly used by your astronomers than the light-year or mile, so I'll be using it to describe distances from here on. In case it makes a difference, and you aren't just reading these values as "blah blah blah *really big number* blah blah," one parsec is a little more than three light-years. And for even more comparison, that's about the same as nineteen trillion miles.

Even though your astronomers use parsecs to quote distances to extragalactic targets, they don't use parallax to measure them. For that, we need to go to the next rung on their distance ladder.

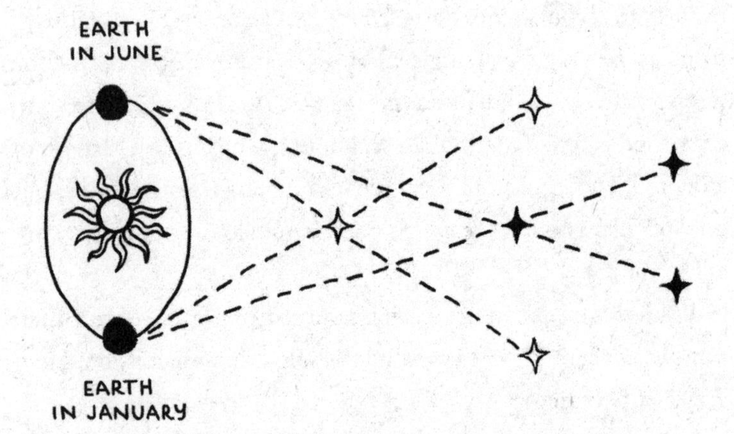
EARTH
IN JUNE
EARTH
IN JANUARY

Back when I was experimenting to see what amusing new shapes I could mold my gas into, as young, rascally galaxies are wont to do, I went through a phase where I played with the luminosity, or brightness, of different objects. I wanted to see how consistently I could produce an object with a given luminosity, so I created some stars that brightened and dimmed over time in a very predictable way. I also made several pairs of stars that would explode with a specific luminosity when their material eventually came in contact with each other. I didn't design any of these stars with you in mind, but your astronomers still use them to measure distances to faraway targets. They call these objects "standard candles."

It's been a few human generations since your kind relied on candles to light up your world, so I'll explain using a lightbulb. Imagine you turn on a lightbulb in a long, dark hallway and start walking away from it. As you get farther away, does the lightbulb appear brighter or dimmer?

I hope to the Cosmos that you said dimmer.

Yes, even though the lightbulb's inherent brightness didn't change, its apparent brightness dwindled as you moved through the imaginary hallway. That happens because as you move away from the lightbulb, its light spreads out over a larger area before it reaches your eyes. Your astronomers and physicists call this the inverse square law. The apparent brightness of an object decreases with the square of your distance from it.

Now you can hopefully understand that if an astronomer knows both the inherent and apparent brightness of some distant object, they can calculate how far away that object is.

The two standard candles your astronomers use most often are Cepheid variable stars and RR Lyrae stars. They work the same way, but Cepheids are brighter than RR Lyrae because they evolve from much more massive stars. Neither one of them should be called a standard candle because they don't have the requisite constant brightness. But human astronomers aren't always the best at naming things. Instead, the stars pulse, getting brighter and dimmer over time in a periodic way. Fortunately for your astronomers who were desperate for a way to measure longer distances, the stars' inherent brightness is related to how quickly they pulse. The faster they pulse, the dimmer they are.

While I'm here, I may as well tell you why they pulse. As the stars are compressed by their own gravity, they become more opaque, or less transparent. The particles that are trapped inside the newly opaque surface heat up, which increases the outward pressure of the gas in the star. The star then expands, becomes more transparent, and cools down as the photons are allowed to escape. This causes the star to contract and starts the cycle all over again. And because hotter stars are brighter, this pulse in size and temperature also causes a predictable pulse in luminosity.

The first variable star was discovered in your eighteenth century, but it took another 150 years for someone to realize the period of a star's variability was related to its brightness. In the early 1900s, a tragically underemployed woman named Henrietta Leavitt made that discovery while working as one of the so-called Harvard computers. Believe it or not, that was the more polite nickname given to the dozens of women who analyzed data at the Harvard College Observatory.[6] Leavitt was observing some of

Larry's Cepheid stars to measure their brightness and noticed that they had a peculiar pulse. Before you start thinking this makes Larry interesting, any of my satellite galaxies would have worked in Larry's place, since they all provide samples of stars that are more or less the same distance from you (the size of any dwarf galaxy is much smaller than the distance between us). Larry just provided a conveniently large sample of equidistant stars for Leavitt to use.

Leavitt's work unlocked the door to greater cosmic distances and helped human astronomers understand the true extent of my magnificence. Cepheids as standard candles were used to prove the existence of other galaxies and even the expansion of the universe. It's safe to say that the course of human astronomy would have looked very different if not for Henrietta Leavitt, and yet no one thought to appropriately award her work until after she died. You fools.

Cepheids aren't the only stars your astronomers use to measure distance. Sometimes they use something called a type 1a supernova as a standard candle. A supernova is a stellar explosion, and *of course* astronomers have separated them into categories. How utterly human. Seriously, no other animals take it as far as you do.

A type 1a supernova is specifically an explosion caused by the merger or exchange of material between two stars that have been orbiting closer and closer to each other over millions of years. At least one of the stars involved in the collision must be a white dwarf—a tiny cooling remnant of a much larger star after it has stopped fusing its hydrogen into helium. Your own sun will become a white dwarf one day! But you'll be dead long before you'd get to witness it.

It's important that a white dwarf is involved because these form from stars that juuuust miss the mass cutoff to explode in a supernova when they finish burning their hydrogen. Instead of exploding, they just get incredibly dense, dense enough that electrons in the star's atoms get pushed too close together. This triggers a phenomenon called electron degeneracy pressure, which pushes outward against the force of gravity that's trying to crush the star in. Theoretically, the electron pressure and gravity could stay in balance forever. But if a white dwarf can accrete, or gather, material from a companion star, it will increase its mass above that supernova threshold and the whole system will explode.

Some of your fellow humans might try to convince you that the explosion happens because the star reaches the so-called Chandrasekhar limit, above which gravity wins and the star collapses under its own weight. That limit is 1.4 times the mass of your sun. However, the humans who say this are wrong, and you should believe me because I'm the one who built these systems in the first place.

You should also trust me because some type 1a supernovae involve white dwarfs with masses above Chandrasekhar's limit. They're called super-Chandras.

The explosion happens at around 1.4 solar masses because that's when a star starts to have enough pressure to fuse carbon in its core. And when that burning carbon meets the oxygen that's so abundant in the rest of the star... BOOM!

I spent billions of years getting the perfect balance of mass and chemical composition for each star in these explosive binary pairs. It was...true artistry. And you people just use it to calculate a distance value.

Since these types of explosions don't necessarily happen right at the Chandrasekhar limit, their luminosities aren't predictable, and they aren't technically standard candles. BUT the supernova's luminosity is related to how quickly the explosion dims out. Brighter supernovae get dim faster. That relationship can be used to turn type 1a supernovae into standard candles.

Before standard candles could be trusted to provide accurate distances, though, they needed to be compared to distances calculated with parallax. This overlap is necessary between every rung of the distance ladder to ensure that each new method actually works.

Using these standard candles, your astronomers were able to track and map more of my dwarf galaxy neighbors, the ones too insignificant to my own existence to bother telling you about: Phoenix, Carina, Sculptor, and dozens of others. Doing so gave your astronomers a better understanding of the shape of our Local Group.

There are also nonstandard candles, of course. These are objects that don't belong to any special constant-luminosity class, but astronomers have figured out how bright they are anyway, like distant galaxies and energetic black holes. Often, the luminosity measurement relies on models, which can introduce uncertainty to the equation. But astronomers are used to being met with uncertainty. Then they spend the rest of their careers trying to drive those error bars down.

Under special circumstances, your astronomers can also use something they call "standard sirens" to determine distances. On the surface, these work in much the same way as the candles—comparing a target's observed quantity to its inherent properties. The sirens used here are gravitational wave sources—events so energetic that they ripple the very fabric of space-time that we're all sitting on. The sources emit signals at specific frequencies that your astronomers sometimes adorably call a chirp.

Some of these sirens and nonstandard candles have been used to measure distances to objects outside of the Local Group, where you can find other clusters of galaxies, some even larger than ours. The nearest is a group your astronomers call the Virgo Cluster—the next town over, if you will. "Town" is an understatement, as the Virgo Cluster is truly massive. Holding more than thirteen hundred galaxies in its grasp, the Virgo Cluster is the Manhattan to the Local Group's...Cleveland? And even that might be too generous a comparison. But just because Virgo has more galaxies than we do doesn't mean any of them are better than me. I mean, Cleveland had that LeBron James character for a while, and he's apparently the greatest of all time.

Beyond Virgo, there's Fornax, Antlia, Draco, and nearly a hundred other galactic cities that together make up the Virgo Supercluster, so named because the Virgo Cluster sits at its center. The universe, after all, is fractal—you can find the same shapes and patterns repeated on larger and larger scales, from atoms to planetary systems to galaxy superclusters.

I've never left the Local Group because I have everything I need here. And honestly, I'm the glue that holds this group together.

Besides, if the universe keeps expanding at its current rate, Virgo and the other clusters will disappear out of sight eventually. Right now, our neighborhoods can feel each other gravitationally, but we are by no means bound to each other.

Your astronomers weren't satisfied with limiting their understanding of the cosmos to the Virgo Supercluster, and nor should they be. It's one of the reasons I continue to hold any kind of respect at all for your species.

To study targets outside of the Virgo Supercluster, your astronomers have to determine how far away they are. And the first few rungs on the distance ladder won't work for objects so far away. The next rung is the standard ruler. I trust you've picked up the pattern by now and know that standard rulers are objects whose physical size can be determined and compared to their observed size to calculate distance.

It's often not individual objects that are used as standard rulers. Sure, sometimes your astronomers try to use individual galaxies in this way, but we're majestic creatures, not mindless sheep who will grow to a certain size just because other galaxies are doing it. Instead, astronomers often use baryonic acoustic oscillations (BAO) as their rulers.

Hmm, you have no idea what that means, do you? I truly don't think any less of you for that. The BAO isn't something you can see with your sad little human eyes, and it's not like you had a choice in that. To explain, I'll have to zoom way out from your rocky home.

Gravity may be weak—even you can temporarily overcome the gravitational pull of your entire planet by flexing a few muscles—but it is relentless. Given enough time, gravity gathers most galaxies into clusters, clusters into superclusters, and superclusters into vast filaments of matter that your astronomers call the cosmic web. There are a few galaxies unlucky enough to be caught in the voids of this web, like Macy, or as your astronomers call it, MCG+01-02-015.[7] With no neighbors within thirty million parsecs, Macy might be the loneliest galaxy in the universe. I'm sure that kind of isolation comes with some disadvantages, but oh, what I wouldn't give to spend a few thousand millennia free from my satellites' expectations, away from the gas-gulping rat race and the pressure to fight other galaxies to survive.

But we don't have to zoom out quite as far as the cosmic web to see that there are over- and underdensities of matter in the universe. These peaks and troughs occur at regular intervals through space, almost as if some powerful ripple worked its way through the universe and collected all the regular matter at the crest of each wave. Indeed, when the universe was young and small and hot enough that atoms hadn't yet formed, gravity tried to clump all the matter together. The extra heat from the densely packed particles produced an outward pressure, and the universe was caught for a time in a harmonic dance between the opposing forces. All that back-and-forth created waves in the matter of the universe, and the shape held as we stretched apart.

Your astronomers knew the BAO was there thanks to a sensitive survey of galaxies throughout the universe called the Sloan Digital Sky Survey (SDSS). Over twenty years and using more than twelve thousand aluminum disks[8] to collect their data, this survey measured the spectra of four million galaxies, black holes, and stars. Thanks to SDSS, your astronomers now have their most accurate 3D map of the galaxies in the universe yet.

The ripples left by the baryonic acoustic oscillation were one of the first discoveries credited to the survey in its first phase, and now your astronomers have measured the BAO to within a 1 percent error. Once they could see the overdensities, they could use the fixed distances between them as standard rulers. The distances gleaned from standard rulers are so large that they helped your astronomers better understand exactly how fast our universe is expanding.

That very same expansion leads to the final rung of the distance ladder: cosmic redshift.

Redshift—and its opposite effect, blueshift—can be seen on smaller scales. It can be heard, too, as this Doppler effect works for any signal that travels as a wave. When a source of light moves away from you, it continues to emit light, but that same light takes just a little bit longer to reach you because it's coming from farther away. You may have heard some of your fellow humans describe this as the light waves getting stretched, but that's not entirely accurate. There's not some force acting directly on the light waves. Each subsequent wave just travels a little farther to reach you as its source recedes. Your scientists call this effect redshift because red light has longer wavelengths than blue light. If you flip all of this around so that the source is moving towards you and the light waves appear to be compressed, you get the effect your scientists call blueshift.

Cosmic redshift works similarly, but in this case, there actually is something stretching the light waves. The expansion of the universe doesn't mean that galaxies are just moving away from me through static surroundings. No, the very space between us is growing, like the surface of a balloon or spreading pancake batter. Light waves caught in that space on their way to you or me get stretched, too.

Your astronomers can measure how much a distant object's light wave has stretched or been shifted towards the red end of the electromagnetic spectrum and use that value to figure out how far away it must be.

Utilizing cosmic redshifts, your astronomers can measure distances to objects near the edge of the observable universe. The sky isn't even the limit anymore; it's the sensitivity of your telescopes. Distant objects appear fainter than closer ones, and fainter objects are harder to observe with your telescopes. Still, your astronomers have pushed the available technology to that limit and observed galaxies at redshifts greater than eleven. The number itself is divorced from any tangible meaning. It's used to scale other values. This distant galaxy, which your astronomers have so creatively dubbed GN-z11—I've never had the pleasure of meeting this galaxy, so I don't know what name it prefers—is 13.4 billion light-years away. That's correct. Billion. Your astronomers are observing GN-z11 as it was only four hundred million years after the Big Bang.

I will say this about those astronomers: they've done a lot with the little they were given. What is it you humans say? Make lemonade out of lemons? Well, in this scenario, all I did was show them a picture of a lemon tree and they figured out the rest.

When—or if—you learned about the scientific method as a

child, you likely learned that it requires experiments. But the truth, human, is that some sciences are observational in nature, not experimental.

Your astronomers haven't yet determined how to build miniature stars to experiment on. They can't create groups of different galaxies and manipulate one to see how it affects their results. Instead, they have to work with what already exists. They search for targets they can assign to control groups, and create test groups out of objects that have been manipulated by nature.

This same type of observational science is even applicable in your own life. I've seen the way most humans try to solve their problems. You use force or trickery to make things go your way. Or you ignore your difficulties and hope they don't come back to—how do you say?—bite you in the ass. (Shouldn't it be *on* the ass?) It would be so much better if you spent more time observing the world around you, keeping an eye out for the root causes of your problems and learning from how other people have gotten themselves out of similar predicaments. That is, after all, how astronomers expanded their horizons to the edge of the observable universe.

Speaking of the edge of the universe, when astronomers do endeavor to convey their work to the rest of you, it's clear the vastness of space makes most of you uncomfortable. Hearing about galaxies billions of light-years away reminds you of your insignificance. And instead of feeling encouraged by your meaninglessness, you let yourself feel disheartened.

Yes, in the grand scheme of things, your life is pointless. You'll never travel to the other side of me or influence anything on the other end of the Local Group. Let's be honest, I'm only talking to you now because I miss hearing the stories of your ancestors and yearn for the brief satisfaction that will come if you start telling them again. I say "brief" because I know your entire species will be gone before too long.

But that means that every human you know—and everyone you don't—is insignificant, too. You are just as important as the celebrities and politicians and influencers who keep your world moving, which is to say "not very." No decision you ever make will have a significant impact on the universe. Isn't that so freeing, to know that your actions don't matter? Maybe they matter to you and your fellow humans, but I guarantee that even on your small scales, most of your choices aren't as meaningful as you fear they are.

I wish I could live a life as inconsequential as yours, unimpeded by something as cumbersome as real responsibility. Alas, true freedom has always eluded me.

CHAPTER SIX

BODY

I MAY HAVE BEEN TOO humble before when I said I put all my responsibilities on autopilot. I'd hate for you to walk away from this believing that I don't do any work, because being a galaxy every day is *exhausting*. In addition to holding this *entire* neighborhood of at least fifty galaxies together, I also have my own gas to transport and mold, and over one hundred billion stars to supervise. Luckily for all of us, my body was made to move stars around.

I wouldn't expect you to know what I look like. It's not like you've ever seen all of me at once. Oh, but so many of you think you have! Well, I love to break it to you, but none of those pictures you've ever seen of me are authentic. They're artists' impressions, though they're usually informed by data. The truth is that no human machine has ever left me, and you can't take a picture of a house when you're inside it.

Generally speaking, my body has three different regions: bulge, disk, and halo.

Let's start with the part you're probably the most familiar with. Those artistic drawings you've seen of me do an excellent job of accentuating my disk—the flat part with my signature spiral arms—but they don't show you everything.

It would be so easy if I could tell you that from one edge to the other, my disk measures exactly thirty kiloparsecs. (I trust you haven't forgotten about parsecs. A kiloparsec is one thousand of them.) But I can't tell you that, because I don't have an "edge." No galaxy does. We're all made of dust and gas, which maintains neither its shape nor its volume if allowed to roam free. So even though I'm big and strong and have enough gravity to hold myself together, those particles near the outskirts are always moving. It makes me kind of fuzzy, and I like it.

Your astronomers, on the other hand, find it frustrating that they can't quote an exact size, so they've come up with a couple of hard, quantifiable measurements. Sometimes they calculate a galaxy's scale length, which is the distance from a galaxy's center to where the brightness is only $1/e$ the peak brightness. This assumes that the galaxy is brightest in its center and then gets exponentially dimmer as you move out towards the (blurry!) edge, which is usually a fair assumption. But not always.

You probably think that numbers must look like...well, numbers, don't you? I keep forgetting how unschooled most of you humans are. But I'm sure you've heard of π and know that it represents a specific number. Similarly, the number e, also known as Euler's number after a Swiss mathematician who was born *after* the number was discovered, is approximately 2.72. Just like π, you'll find e everywhere in nature if you bother to look, from the compounded interest in your bank account to the probability of winning a game of random chance.

Astronomers also sometimes calculate a galaxy's half-light radius (the radius where the brightness decreases by half from the peak) and its half-mass radius (surely I don't have to explain that one to you).

But, if you're willing to accept a little bit of fuzziness, my disk has a radius of about 15 kiloparsecs, which I'll shorten to kpc from here on. Your little solar system is about 8 kpc from my edge, so you're like the most average system there is. Congratulations on your mediocrity!

You probably think of the disk as flat, which is technically accurate because the entire universe is flat, but it's not *precise*. In actuality, my disk is about 1 kpc thick from top to bottom. Your planet is, again, pretty much smack-dab in the middle of the vertical axis, though you're ever so slightly above the midplane.

My disk holds 70–85 percent of my stars—they can move in and out of my different regions as they follow their orbits—and it's where I make most of my new ones. The disk stars are the most well behaved, moving on nice, circular orbits around my center. Not perfectly circular, of course. There's some deviation that your astronomers call epicycles, and they kind of look like... what's it called? A Slinky! Disk star orbits look like a phenomenally long Slinky stretched out to form a circle going around my center, and the stars follow the curves as they orbit.

Their circular orbits make them so much easier to keep track of than my other stars because I can predict where they'll be and when. I'm so terribly busy that it's a relief to know that I can look away from a disk star for a few million years without worrying about where it might wander off to. And that's not a fluke, it's just how disks move. A little trick of the galaxy trade, if you will. The best human pizza chefs know this trick, too. Balls of pizza dough and giant clouds of gas alike tend to flatten when they spin as most of the material collapses to the center and centrifugal acceleration stretches it radially in the plane of rotation.

Because disks are much thinner than they are wide, most of the matter is concentrated around a flat plane. Having so little material above or below the plane means that gravity mostly operates in only two directions: towards and away from my center, which is the only place anyone should want to go anyway. The little material that does exist away from the plane is what causes those epicycles I mentioned before. But I can't let all the stars go to my center because it would create such a jumbled mess, so I gave them a little bit of a twist, literally.

I keep my disk constantly rotating so my gas is spinning along with me, which prevents it from falling towards the middle. Your scientists call this the conservation of angular momentum. If you have a rotating object, like a galactic disk or a human figure skater, I guess, and you make it smaller, it needs to spin faster. For a star, this means that if it moves to a smaller radius, closer to the center, it also has to orbit faster, and who has that kind of energy to spend? Not my stars, though sometimes they'll trade angular momentum with each other so they can switch orbits. But for the most part, everyone stays in their lane. So the shape of my disk is supported by rotation because I put in the work to make it that way. I'm exceptionally good at what I do.

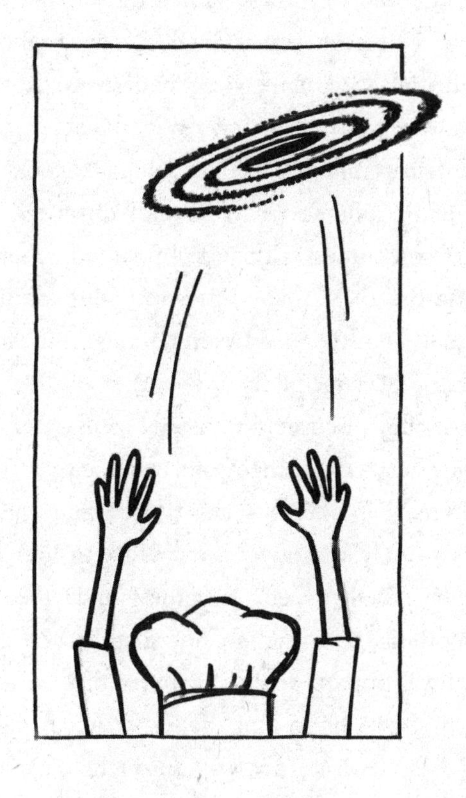

Speaking of rotating, you're doubtless wondering how I got those stunning spiral arms you've seen in all the pictures. I have two big ones that attach to my central bar and swirl all the way around my body. Your astronomers have named them Perseus and Scutum-Centaurus. *I* have named them…nothing. Because they're arms. Not even you strange humans name your arms, even though you always name your children. Weird.

I was, however, so tickled when you named one of them Scutum-Centaurus that I've taken to calling it Scoot.

Perseus and Scoot also have several offshoots, which your astronomers sometimes call spurs. Apparently, their names are Carina-Sagittarius (how many times can your scientists possibly use this name?!), Norma, and Orion-Cygnus. Your solar system is nestled right on the edge of my Orion arm.

As soon as they discovered my spiral arms, your astronomers started asking what caused them, and over the years, they've come up with two real hypotheses. That's right, even I, a glorious celestial body who only recently started utilizing human language, know the difference between a hypothesis and a theory.

One hypothesis says, though not so theatrically, that in the beginning, I was but an evenly distributed disk until some of my gas clumped, gathering in long, dense streaks fanning out from my middle. As I twirled my way through the universe, those streaks got caught up and twisted along with me. The problem with this little supposition is that my arms would be more tightly wound than they are, since I've spun around so many times. Your own sun orbits my center once every 250 million years. At only 4.5 billion years old, you've gone around my whole disk roughly eighteen times. Add another forty rotations to that, and you can just imagine how tight my arms would be.

The other hypothesis is that my spiral arms aren't material arms at all, meaning they're not like ropes made of gas and stars that move together as I drag them. Instead, my arms are like traffic jams, but rather than being caused by humans' trademark slow reaction speeds, they're caused by a wave of density moving through my whole body. Stars and dust and gas move more slowly when they get caught up in the wave, so material is bunched up and the area grows denser, but they still move on. Eventually, everything in my disk will pass through the wave.

For a while, your astronomers thought this density wave hypothesis couldn't be accurate because density waves should be relatively short-lived—the wave should move through after only a couple billion years and the spiral pattern should wash out. But they were underestimating the strength of my gravity. The force applied by the material surrounding my arms holds them in shape. It's like a sausage, but less gross and more impressive.

There are a few different ways to create these density waves. Every spiral galaxy has their favorite method. For example, the Whirlpool Galaxy, who lives in another galaxy cluster, prefers the tidal method. This involves asking a few companion galaxies—probably dwarfs—to orbit by and drag some gas into an arc.

The galaxy your astronomers call NGC 1300 prefers the central bar method. This takes more work on the galaxy's part, but it does create more symmetrical spirals, and every beautiful creature knows that looking good requires effort. A galactic bar is a co-moving block of stars in the center of a galaxy. It usually houses the galaxy's supermassive black hole and a dense collection of stars, so it carries a lot of mass,[1] which means it has a strong gravitational pull. As the bar rotates, it adds a resonant boost to some of the stars in the disk. Resonance means that the orbital periods of the two bodies involved are integer multiples of each other. For every one orbit of star A, star B orbits exactly twice, or thrice, or…you don't have a word like that for the fourth time. On Earth, you may have experienced this if you ever pushed someone on a swing or hit a punching bag. If you make contact at exactly the right moment, you can add to the height of the swing.

A galaxy's bar is like your fist, and its stars are that bag. If a star is in resonance with a bar, it gets an extra boost in speed every time its orbit lines up with the bar's. This happens all the way out to the galaxy's fuzzy edge, but because material closer to the center of the galaxy often moves at least a bit faster than the outer stuff, the inner material is moving faster than the bar and the outer material is moving slower. So, the inner material gets pushed ahead of the bar, and the stuff by the edge lags behind. That's how you get those lovely, symmetrical spirals that make everyone else jealous.

But back to more important things: me. Sometimes the spiral arms make my disk stars act up. They occasionally use the "traffic jam," so to speak, as an excuse to move to different radii without my knowing. This means I lose track of them. I put up with it, though, because my spiral arms are the best in the neighborhood, and I'll do anything to maintain that position of superiority. Besides, they don't require anywhere near as much attention as my bulge stars.

In my center is my galactic bulge—no, not that way, you anthropocentric sicko. My bulge is a dense, chaotic, mostly spherical collection of stars in the middle of everything. Well, not *everything* everything because even I'm not narcissistic enough to believe I'm actually the center of the universe. There is no center of the universe. But the bulge is where my bar is; it's where 15 percent of my mass is: stars, gas, dust, *and* dark matter; it's where Sarge, my supermassive black hole, lives. Even though those bulge stars are huge pains in my metaphorical ass, my bulge is vital to me.

The bulge is small compared to the rest of my body. Remember my disk is 30 kpc from edge to edge? The bulge is just 2 kpc across in all directions. But because it is nearly spherical, gravity works in three dimensions instead of just one, so orbits are so much more complicated. Some stars orbit in ellipses, which is at least merely a stretched-out circle. That's not too hard to keep track of. But other stars' orbits trace out these wild rosette shapes, and some of them move in something kind of like a figure eight, and they all require so much focus!

Still, my bulge is an interesting part of my body because it's where some of my oldest stars are, those from the original little protogalaxies that combined to form me. It's also where some of the most exciting action happens. And I think you'll agree because it all relates to your incessant search for extraterrestrial life.

Most biological functions are affected by high amounts of the most energetic radiation, like those X- and gamma rays that you can't see. The most dangerous sources of these types of radiation are supernova explosions. Yes, human, like the explosions that your astronomers use as standard candles to calculate distances. You're learning! Except it's not just the type 1a supernovae that give off radiation, it's all kinds of supernovae: the ones that happen when

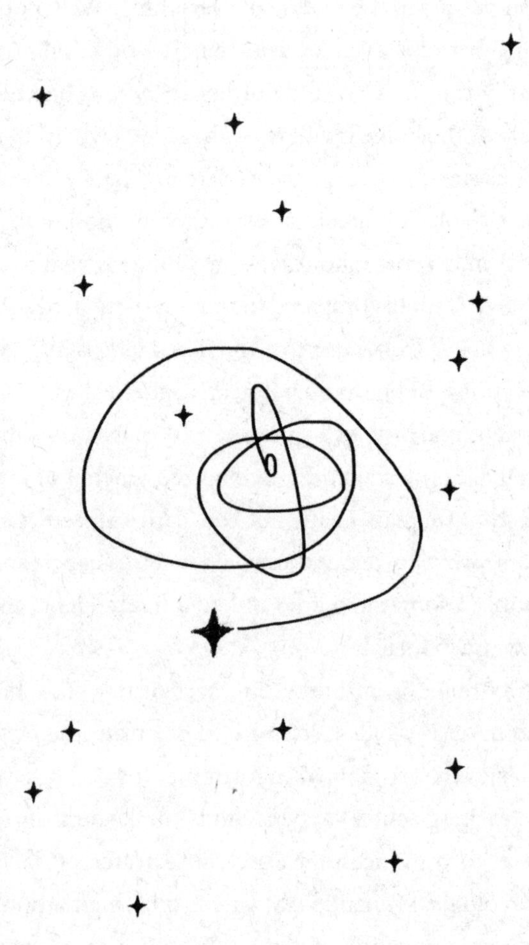

massive stars burn through all of their core hydrogen and the ones that happen when a white dwarf gets a little handsy with a bigger star. Your planet is so fragile that it would become uninhabitable if a supernova went off within fifteen parsecs of you. That kind of energy would rip your planet's protective ozone layer to shreds. It would ionize your atmosphere, essentially stripping electrons away so that the thin layer of gas between you and the cold vacuum of space is suddenly made of weird, charged particles, and disrupt the photosynthesis that underpins your planet's whole food chain. Luckily for you, there are no stars in imminent danger of going supernova anywhere nearby. But the same can't be said for stars in the bulge, where things are much closer together.

That same closeness means that bulge stars are more likely to have close gravitational interactions with each other. In fact, most of my bulge stars—about 80 percent of them—come within 1,000 astronomical units, or AU, of another star every billion years.[2] I've just introduced a new unit that your astronomers use, so 1 AU is the distance between you and your sun. It's some negligible number... about ninety-three million miles, I think? Anyway, if a star came within 1,000 AU of your sun, it would be passing through your solar system.

These close stellar encounters can cause so much drama! Sometimes the stars rip other stars' planets away. Sometimes they get sneaky and only tug on another star's planet just enough to make it fly out of its system a few million years later. By then, the interloping star is too long gone to face any of the consequences. A star can even use one of these gravitational encounters to prevent another star from forming its planets in the first place.

This type of ruthless environment isn't the most hospitable for soft creatures like you.

A few of your astronomers—yes, disappointingly, it really is that small a field—are interested in this question of habitability in different parts of my body. So far, they've concluded that my bulge isn't an ideal place to focus their search, and that the best place for life is where you already are. You're even at the right distance from my center to be orbiting me at the same speed as my spiral arms, so you don't have to worry about one catching up with you and your frail planet being exposed to that dense environment.

The biggest but dimmest part of my majestic body is my halo, which has three overlapping parts: stellar, circumgalactic, and dark matter. The stellar component is a messy spherical field of stars and globular clusters[3] left over from previous interactions with other galaxies, and it extends out to about 100 kpc. The circumgalactic halo is a cloud of warm gas that I can use to fuel my star making. The dark matter halo is my most extensive and massive "organ," you could say. It's not called "dark" because it's inherently wicked or threatening. In fact, dark matter was instrumental for all of us in the beginning. Without dark matter's cold clumpiness, galaxies like me wouldn't have been able to hold on to mass and form stars in the early, hot universe. No, it's merely called dark because it doesn't interact with light, and the astronomers who named it weren't very creative. Dark matter doesn't emit, absorb, or reflect any electromagnetic radiation—light. Therefore, it must be made of different material than the stuff you can see. Your astronomers don't know what it is—and I won't let that tidbit of information slip—but they know what it can do.

I've hinted before that gravity is a galaxy's most valuable tool. Well, since dark matter is made of a material that interacts gravitationally, but not electromagnetically, it's like a secret weapon! A tool that can be felt but not seen. And I have *a lot* of it. If you think my disk is big...ha! My dark matter halo extends out to 600 kpc. To give you a sense of its mass, I am 1.5 trillion times more massive than your sun. That's 3×10^{42} kg. If I could somehow fit the entirety of my body on your planet—I've always wanted to experience your own local acceleration due to gravity—I would weigh about 6.5×10^{42} pounds. And 84 percent of all of that is dark matter.

This, by the way, is what your astronomers mean by those $\Omega_{m,\ rel,\ \Lambda}$ symbols you may have seen if you've done your own reading about the origins of our universe. Those are the relative densities of the universe when compared to the "critical density," the density that, according to their predictive models, means the universe will continue to expand forever. So far, they've found that about 68 percent of the matter/energy (because the two are interchangeable when you reach galaxy brain status) in the universe is dark energy, a rather imprecise catchall term for the as-yet-unidentified-by-humans force that's pushing the expansion of the universe. Another 27 percent of the universe is made of dark matter, and about 5 percent is regular baryonic matter like you. A teeny tiny fraction is made up of relativistic particles that move at or near the speed of light and carry electromagnetic energy. It's all adding up to a number dangerously close to that critical density, but I'm not at the part of the story where I tell you what will happen if the universe just keeps expanding forever.

It's fortunate that dark matter makes up so much of the

universe, and that I got so much of it (let's all pour one out for those sad little dark matter–poor galaxies[4]), because my halo is the reason I've made it this far. Back in the beginning, a few hundred million years after the Big Bang, when the first protogalaxies were being born, the universe was too hot for all that non-dark matter, or the baryonic matter as your scientists call it, to bind together gravitationally. Gas particles were moving around so fast they would have overcome the gravitational attraction of the other baryons. But what if there was already some cooler material that clumped more easily and could still attract the warmer particles? Dark matter is like a scaffold you might build to help your plants grow. So we all owe our existence to that dark matter.

But dark matter doesn't come without downsides.

As we know because of the conservation of angular momentum, the stars farther out in a galaxy's disk should move more slowly than close-in stars. But that simple relation only works if most of the mass is concentrated in the middle. Since my dark matter halo is so huge and holds most of my mass, it affects the speeds of my outer stars. They move faster than they would if they were just reacting to the mass of the luminous material...the shiny stuff. They move so fast that some of them could even have escaped if I didn't have as much dark matter as I do, holding them in place. The slope of that radius-speed relation, aka the rotation curve, depends on the amount of dark matter the galaxy has, and it's how your astronomers learned about dark matter in the first place.

Your physicists first hypothesized dark matter in 1933, but you didn't find evidence until 1968, when Vera Rubin realized stars were moving faster than she expected. Rubin was looking at the rotation speeds of stars only because she wanted to study something uncontroversial after her previous work on more contentious topics was mocked or ignored by her contemporaries.[5] The amount of human knowledge that's been prevented by your species' asinine ideas about what kind of human deserves notice is truly staggering. It's almost as ridiculous as the idea that there are different kinds of humans in the first place. Despite the host of obstacles placed in her way, Rubin ended up discovering game-changing astronomical evidence. She found the same trend in other galaxies, proving that it wasn't a fluke. This gutsy pioneer was the first one of you to get over your visual bias and finally know me, all of me. And it still took you until 2020 to name a telescope after her. Show some more respect next time.

The more dark matter a galaxy has, the faster its outer stars can rotate, so the "flatter" the rotation curve. But if a galaxy doesn't have much dark matter, there's not as much mass to speed up the stars, so the rotation curve has a downward slope. When Trin said my rotation curve was looking flat, that was a diss on my weight, as if I'm not proud of the dark matter I've packed on. But what can I say? Trin's a petty bitch.

Vera Rubin's work wasn't the first monumental shift in humanity's understanding of my body. Aristotle saw the streak of my disk flashing across your night sky three hundred years before the birth of your Christ and called me Galaxias, coming from the old Greek word for "milk." It's where you get your modern word "galaxy." Looking upon the little slice of me that he could see, Aristotle believed it to be the point where the terrestrial and celestial spheres met and ignited to create a perpetual flame. A little more than one thousand years later, a man known to some as Ibn Bājja hypothesized that the band of light across the sky was in fact a distant collection of stars packed tightly together. That hypothesis was confirmed in 1610 when that Galileo guy looked through a telescope and resolved the individual stars in a strip across your sky. My disk stars.

Once humans accepted the notion that I was a collection of stars, they started wondering about my shape. In 1750, a human named Thomas Wright proposed that I was arranged in a flat layer.

Soon after, in 1785, Caroline Herschel and her brother published the first-ever methodically created human-made map of my body. They did it by mapping the stars you can see from Earth, but the method they used to find the stars' locations was faulty. It relied on the absurd assumptions that my stars are distributed uniformly throughout my body and that their devices could see every one of my stellar creations. How presumptuous! I bet the map would have been better if Caroline hadn't had to waste her time spoon-feeding soup to her perfectly capable brother while he "toiled away."[6] It looks like a real Larry and Sammy situation to me, one partner obviously contributing more than the other. It's a shame you humans usually can't choose your siblings. But then again, Sammy chose Larry, so I guess love is a mystery to us all.

By the twentieth century, the consensus among human astronomers was that my body was a flat disk about twenty thousand light-years across (the word "parsec" was just about to be invented in 1913 by an astronomer named Frank Dyson[7]) and that your sun sat somewhere near the middle. Will you humans ever miss an opportunity to try to put yourselves at the center of everything?

A young rebel named Harlow Shapley had been using Cepheid variables (thanks to Henrietta Leavitt) to map globular clusters in my halo. Globular clusters are just what your astronomers call collections of gas, dust, and up to thousands of stars that are gravitationally bound together within a galaxy. Shapley drew two conclusions from his work. The first was that my body was much larger than his colleagues thought, closer to three hundred thousand light-years, or 90 kpc. He was wrong, of course, but he redeemed himself with his second result: the sun is closer to the edge of the Milky Way's disk than its center. Finally!

Shapley had noticed that most of the distant "nebulae" were clustered in one direction, towards the constellation Sagittarius, and not uniformly distributed, as was the belief of the astronomy old guard. Unfortunately, he reached his accurate conclusion only through faulty reasoning. He thought that the nebulae must be part of me, because in his head, I was far too large for there to be anything outside of my bounds. A more established astronomer named Heber Curtis disagreed with most of what Shapley said, especially the part about my being big enough to encompass the nebulae, which he believed were their own "island universes" just like me.

In 1920, Shapley and Curtis were invited by the National Academy of Sciences in the US to publicly defend their respective views on the architecture of the universe. A whole room of people gathered to discuss my position! Your astronomers call it the Great Debate, and I was flattered.

Around the same time as the debate, Edwin Hubble, a newly minted astronomy PhD, was using Cepheid variables to determine the distances to dim nebulae. In 1924, he ended the debate once and for all when he found that the variable stars in what he knew as the Andromeda nebula were much too far away to be in me, thus confirming the existence of other galaxies.

A couple of years later, Hubble published his Hubble sequence of galaxies, also known as the Hubble tuning fork, based on his observations of other galaxies. The audacity! Until then, I thought Hubble was surprisingly acceptable for a human because he added to your collective knowledge about me. He discovered that I'm not the only galaxy in the universe and then was obnoxious enough to immediately begin categorizing us and predicting our behavior based on our shapes. I suppose it would be unfair to remain angry with him because he was right. For galaxies, shape is an important characteristic that hints at our past and informs our future.

But I can hold a grudge against him for being so judgy about it! I really didn't appreciate a human saying I have to act a certain way just because of how I look. Save that prescriptive anthropoid bullshit for your own planet. I've lived and worked in my body for billions of years, so I know what I look like. I've even spent plenty of millennia comparing my body to those of the galaxies around me, usually coming away quite pleased with how I measured up. But to pigeonhole us like that? I've heard some of you describe bodies as "pears" or "hourglasses," and I bet you wouldn't like it if I started predicting your behavior based on your figure.

The tuning fork separated us galaxies into spirals like me on the right and elliptical galaxies on the left, with intermediate steps in between. Elliptical galaxies are ellipsoidal, much like a bigger version of my bulge, so they don't have any strong features like spiral arms.

Hubble dubbed ellipticals "early-type galaxies" and spirals "late types," which is the opposite of fact. Elliptical galaxies are often made of smaller spiral galaxies that collided at inopportune angles.

Elliptical galaxies have more old stars and less star formation than spirals, so they're cooler and dimmer. They're not common in the Virgo Supercluster—that giant cluster of galaxies we're all a part of—but they are more frequent at the centers of dense galaxy clusters than near the edges. And these poor galaxies, bless their hearts, have no disks, so all their stars orbit like my bulge stars. I don't envy them the chaos.

Though I should keep in mind that we'll all be like that one day. Well, I will, at least, since you'll all be long gone.

Hubble also categorized spiral galaxies by how tightly their spirals were wound. And between spirals and ellipticals sat lenticulars: galaxies with a large central bulge and an extended disk with no spiral arms.

These days, in addition to grouping by shape, your astronomers also organize galaxies according to their size, luminosity, star formation rate, and the strength of their central black hole. And if your scientists see a galaxy that really doesn't fit neatly into their little classification scheme, they call it "irregular."

Despite coming so far in your galactic taxonomy since Hubble's tuning fork, you went and named a telescope after him! It launched into space in 1990, and it's just as captious as its namesake. I'll admit it does good work, though. Because of that telescope, your astronomers have discovered that there are hundreds of billions of galaxies just in the part of the universe close enough for us to see. They've used it to get more precise measurements for the speed of the universe's expansion and even to look at what might be the first moon your kind has ever discovered outside of your solar system.[8]

The success of the Hubble telescope paved the way for larger and more advanced observatories, both on ground and in what passes for "outer space" on your planet, which is only a few hundred miles off your surface.

The Kepler telescope you launched in 2009 opened your eyes to the billions of planets outside of your solar system. In thirty years, you've discovered nearly five thousand of these exoplanets. Your astronomers claim they haven't found more because it's so hard to see them. You know what's truly difficult? Creating, tracking, and inventorying all the planets in my body, including the many billions that you've never seen. But hey, I'm all for anything that gets you closer to really knowing me.

In 2013, you launched a telescope named after an ancient goddess who personified the Earth. Your Gaia, in her latest incarnation, has yielded the most precise map of my stars your kind has ever created. It includes more than a billion targets, and many of them bulge stars, which you humans struggle to see because there's so much dust in the way. Gaia even tracked their motions, so now your astronomers can infer a star's entire orbit, even if it would take 250 million years to complete like your sun's. If I had breath, it would have been taken away when I saw myself rendered so completely on your screens for the first time.

I was so thrilled that I can even admit that it's not your fault that you live too close to my midplane to see all of me clearly. It's not my fault, either! I didn't put you there on purpose. Believe me, I have more important things to worry about than which planets get the unobstructed view. Your world was just unlucky, so your astronomers had to try harder. And their assiduous work led them to MeerKAT.

If the Hubble telescope made me feel like I was being watched by some creep with a Polaroid, then your MeerKAT telescope in South Africa turned me into a professional model posing for the most sought-after photographer. MeerKAT has provided so many stunning images, especially of the glorious gas in my bulge, close to Sarge in the center.

MeerKAT can see me so clearly because it looks at the right wavelength to see through the dust between you and my bulge. Radio waves cut right past the pesky stuff. MeerKAT also isn't just one telescope; it's sixty-four telescopes working together. This method is known on your planet as interferometry because it relies on studying interference patterns of light waves from different sources. With it, your astronomers have built networks of telescopes as big as Earth itself to take a picture of another galaxy's supermassive black hole. You didn't get mine, though. Yet...

MeerKAT reminded me of something we should all recognize: I'm fly as hell! I'm beautiful and strong and I do my job well. It's not like I didn't already know that, but those images were the tributes I needed after taking a few blows. And I'm not just referring to your entire species forgetting about me for the last three hundred years.

CHAPTER SEVEN

MODERN MYTHS

OKAY, OKAY, YOU DIDN'T *ALL* forget about me these last three centuries. My value continues to be acknowledged by your astronomers (both those who study my kind in offices for money and those who do it in their yards for fun), the astrologers (you might think they would annoy me, but I like them; they remind me of your sufficiently awestruck ancestors[1]), and of course the sci-fi nerds.

From what I can tell, "nerd" can either be a compliment or an insult. I mostly mean it as the former because believe it or not, it's these nerds that have kept the tradition of mythologizing space alive.

Did you think myths were just things of the past? Absolutely not. Humans make new myths every day about the things you want to have faith in, like politicians who keep their promises or the idea of the altruistic billionaire. At their core, myths are stories that you capital *B* Believe in, even if you know some (or all) of it to be false. What does truth mean, anyway, when it butts up against a narrative that you've folded into your identity? After witnessing more than eighty years' worth of science fiction conventions, I'm convinced that your beloved sci-fi stories should certainly be considered myths.

Take for example the myth of the interstellar political coalition, the idea that there's a vast network of planets that operate according to standardized rules. In most versions of this myth, an alien network is already out there, and they make contact with humanity once you become technologically advanced enough, usually after you develop faster-than-light travel. As a species, you have zero evidence that any of this is even possible, yet you've told this story more times than a mere human could count because you Believe in it. You want it to be true so badly that you're rushing to get people off your little rock. I suppose I'd want to do the same if I were trapped on the same planet my whole life, but I hope for your sakes that the rushing doesn't lead to too many harmful mistakes.

Let's not forget the myth of the united human race. Much of your space-based science fiction takes place in a future where both the pet peeves and deep injustices of your kind have been rectified and smoothed over. Even in the 1960s—a time when darker-skinned humans were barred from most positions of power—*Star Trek* dared to envision a future where a Black human woman could work as an officer aboard a federation ship.

In your mythical sci-fi universes, Earth as part of a galactic federation is an Earth where the indelible parts of a person's identity have no bearing on their life. After all, what does a difference in silly chemicals like melanin or estrogen matter when you're face-to-whatever with an alien who couldn't point to Earth on one of your Gaia-made maps? (Actually, I'm not sure many humans could point to Earth on a map of my body, either, but you get the gist.)

Unlike most of the space myths of yore, the ones you tell today aren't meant to explain how things *are*, but instead they show how you *wish* things could become. Science fiction stories are aspirational myths. They're humanity's dreams for the future. And of course, those dreams were inspired by me, the one sky that watches over all of you.

As pleased as I am to be a part of and entertained by your new myths—your method of putting your stories on screens instead of sharing them around the fire is much more stimulating—I have some complaints about how I've been portrayed.

First of all, you've turned me from a character into a setting. Instead of being a powerful god watching over you at night, I'm just the matter you fly through in your fancy ships. Hurray for inventing new genres or whatever, but shame on you for leaving the most impressive parts of me out of it.

Also, most of the systems your mythical federations use to describe locations in my body are nonsense. Granted, the golden age of human science fiction came decades before your astronomers created the stunningly accurate maps they have now, but still...

I'll pick on *Star Trek* here, not because it's the worst offender but because it's the franchise that most humans are familiar with. Millions of you have collectively spent billions of dollars to mark your enthusiasm for this franchise. For the rest of you who put your stock in other fictional universes and don't know Picard from Pickle Rick (animation is for everyone, even nearly immortal and omniscient galaxies), *Star Trek* has spent decades exploring every possible kind of interaction between members of an interplanetary federation of species who live scattered throughout my body, which they've delineated into quadrants.

Quadrants! How impractical is that? With a 30 kpc diameter and a height of 1 kpc, I am proudly one of the biggest galaxies in the neighborhood. That means one of my "quadrants" would be nearly 200 cubic kiloparsecs! I guarantee your tiny human brain can't even fathom how big that is, and your *Star Trek* writers expect trained space explorers to use that as some kind of useful location signifier? Saying something is in the "delta quadrant" gives very little information!

Now, I'm not saying your astronomers' galactic coordinate systems are much better, but at least they understand the importance of specificity when you're dealing with spaces as large as me.

Some human astronomers use the galactic latitude-longitude (designated by *b* and *l*, respectively) system. Your planet rotates too much to project your local coordinate grid directly onto my spherical-ish body, so the zero-degree longitude line—the galactic prime meridian, so to speak—passes over the line that goes from your sun to the center of the galaxy, and latitude is an angle measured from my midplane, just like yours is measured from your equator. And because space has three spatial dimensions (plus that fourth time dimension that many of you struggle to grasp), there's also a third coordinate that specifies how far away the object is. That distance is usually measured from my center, although sometimes your astronomers measure it from your sun. And they—to my great frustration—don't always say which reference point they're using. I imagine it must be doubly infuriating for other astronomers who use those measurements in their own work.

Then there's the coordinate system that uses right ascension and declination, which likens your solar system to a bulbous clock face. Right ascension, or RA, is similar to galactic longitude, but it's measured in hours instead of degrees. Declination is essentially just a projection of your latitude as it is measured from your equator.[2] It's genuinely so self-centered of you to use this system. Your sun's position in this system changes because it's based on a perspective where your sun appears to orbit your planet instead of the other way around. That view of the universe made sense with the information your ancestors had available, but you know better by now.

I'm sure these methods are useful to your astronomers and the rest of you humans who have yet to venture outside of your solar system, but they would be useless for galaxy-wide communication. They depend so much on your position. If your species manages to last another one hundred million years without destroying everything you've created, your solar system will have orbited far enough from your current position—practically to the opposite side of my disk—that you'll have to come up with more sensible coordinates.

But no matter how silly the *Star Trek* quadrants are, at least they knew a parsec was a unit of distance, not time. Yes, yes, *Star Wars* tried to correct that mistake. I'm still not satisfied.

My other biggest complaint with human science fiction is your tendency to include mostly humanoid aliens. *Stargate* did this, as did *Farscape*, *Men in Black*, *Hitchhiker's Guide to the Galaxy*, and even *Alien*! The creators of these aliens may change the head shape a little, but anyone watching would start to think all life-forms in the universe had to have a head, four limbs, and a torso just like you. As if your stumbling bipedalism and opaque skin are the only possible evolutionary endgames. You can't even see your internal organs! You're required to sit in special machines just to make sure everything inside is working adequately. I'm not going to tell you whether there's other life out there, but if there *were*, there's no way most of it would look like you.

Of course, most of the time your on-screen aliens look like you because of budget constraints or the fact that non-humanoid characters would be less relatable to a decidedly human audience. Some of the more defensive sci-fi nerds have even tried to explain the proliferation of human features across various fictional mes by saying that an ancient humanoid species spread their DNA early on. That's clever, but ultimately less interesting than if you imagined alien species that suited the characteristics of the worlds they evolved on. But what do I know about alien biology? I just contain every alien world your species could ever hope to interact with. That's all.

I worry that seeing all those familiar-looking aliens on their rocky planets with breathable atmospheres will give you the wrong idea about what lies outside your solar system. I have hundreds of *billions* of planets, each of them a unique combination of features with its own random events that influence any biological evolutionary paths. I have gaseous planets, water worlds, lava worlds, planets that orbit multiple stars, even planets that don't orbit any stars. Think about how divergent each of those worlds could be from your own. It might help you realize the error of limiting your imagination.

Speaking of error, human science fiction writers clearly have no idea what a nebula is. To be fair, until about a hundred years ago, human astronomers used the word "nebula" to refer to any fuzzy patch of light in the night sky. Your astronomers have since learned that nebulae are clouds of gas and dust denser than the space surrounding them. There are lots of ways for nebulae to form. Some of them form when clouds of more diffuse gas cool down and start to condense. Others are areas of active star formation, like my own personal star-making laboratories. (Maybe you've heard of the Orion Nebula? It's the closest of these stellar nurseries to your sun, just a few hundred parsecs away.) Others still are the sites of supernova explosions, the death throes of massive stars.

Popular sci-fi films and TV shows don't really concern themselves with the nuances of nebulae. For some reason beyond my nearly infallible comprehension, most humans don't consider in-depth explanations of galactic fluid dynamics entertaining. Every nebula the various spaceship crews encounter looks like a big, colorful, visible cloud of gas, but your feeble human eyes wouldn't be able to see a real nebula if you found yourself floating smack-dab in the center of one.

Nebulae are denser than the space around them, but that surrounding space is *extremely* diffuse. To give you some point of reference: there are about five particles in every cubic centimeter of deep space. The average nebula has a few thousand particles per cubic centimeter of space. That may seem like a lot until you compare it to your planet's atmosphere, which has about 10^{19} particles per cubic centimeter. Ten. Quintillion. Particles. All squished into a space the size of your fingertip. And that's just air!

Star Trek got that one wrong, too. But with sixty years of content spanning a dozen TV series, as many films, and countless games and comics, they were bound to make a few mistakes. As long as you, human reader, don't interpret those mistakes as truth and instead recognize them as the myths they are, I have no quarrel with you.

Even if the hard science gets a little bit lost to some, it's essential that you humans continue making myths. They motivate you to achieve feats that are currently out of reach, like when Arthur C. Clarke imagined satellite communication a dozen years before you launched Sputnik into Earth's orbit. The myths you've told as science fiction have inspired you to reach for credit cards, the internet, and the International "Space" Station. You'll need to keep telling stories if you hope to last as a species, because your planet's days are numbered.

CHAPTER EIGHT

GROWING PAINS

You've made it this far in my story, so I suppose we know each other well enough that I can get deeply honest about how it's felt to be your galaxy these last thirteen billion years. But don't pat yourself on the back for managing to read half of this book. I'm the one doing the onerous and, frankly, humiliating work of explaining myself to a hopelessly corporeal creature.

If I had to guess, I'd say the hardest thing about being a human is your short life span. Not the dying that happens at the end—no, knowing you're going to die eventually is what gives your short lives meaning—but the fact that you see everything as only a snapshot.

Back when your ancestors used to look to me for guidance, before they had telescopes to see the truth of things, some of them noticed that there were two kinds of stars: wandering and fixed. "Star" is a concept that has changed meaning over time, just like "nebula" and "universe." In this context, a star is any single bright point of light in the night sky. Wandering stars appeared to follow their own paths across the sky according to their own individual patterns, while fixed stars stayed still relative to each other, even as they all revolved around your planet, which many of your ancestors believed to be the center of everything.

In reality, only one of the wandering stars was a star at all, and it was your sun. The others were your moon and the planets your weak human eyes can see without aid. Even your word "planet" comes from the Greek word meaning "wanderer." By most accounts, there were seven wandering stars: sun, moon, Mercury, Venus, Mars, Jupiter, and Saturn (a lucky few humans can see the planet you call Uranus without a telescope, but not enough for their perspective to matter here). The humans of ancient Babylon created the seven-day week that you use today based on these seven bodies.[1]

But I digress. The point I'm trying to make here is that many humans thought the distant stars were fixed, unchanging, because you don't live long enough to see them move. You don't live long enough to see me evolve. And that's frankly appalling for you because it means you lack the perspective necessary to really think about anything beyond yourself and your immediate surroundings, but it's *extremely* unfortunate for me because I don't get to feel *seen*. Or heard. Or appreciated for the work I do.

Sammy and Larry are so wrapped up in each other and the stars they're making together that they don't really pay attention to what I do anymore. Trin just makes everything worse. And Andromeda... well, we'll get to that soon.

Everyone else is gone, carried away by the mysterious force, this *dark energy* as you call it, that's propelling the expansion of the universe.

And that leaves me with you, so you better listen up.

As a galaxy, I must do a couple of things to survive, to grow. I've ripped apart more galaxies than you can imagine to collect gas and to keep myself from being torn apart instead. Like I said, I *had* to do it, but a small part of me also started to like it. To relish the destruction. Anything to break up the monotony of millennia alone, amiright?

I'm also compelled to produce stars. Think about it. A galaxy without—or with little—dark matter is rare enough that it's worthy of note. Human astronomers have even published scientific papers just to tell each other when they've found a new galaxy with less dark matter than they expected. But a galaxy without stars is inconceivable.

So I've made more stars than you could possibly hope to count yourself. In doing so, I used up most of the gas I had when I first formed, which means the stars I make these days aren't the same kinds of stars I used to forge. And in the last thirteen billion years, I've felt—not just watched—so many of my stars die. Stars I made with my own not-hands. It hurt deeply, but I had to keep making them—I *have* to keep making them—because that's the job.

Do you have any idea what it's like to produce something when you *know* for sure it's going to die before you? And that you're going to feel it as it's happening because that thing you created is literally a part of you? No, I can't imagine that you do. I'll just have to describe it in detail.

First, I should say that dying doesn't mean the same thing for

galaxies as it does for humans. I wouldn't even be using the word "dying" at all if your astronomers hadn't already started using it in this context.

The benefit of a near-infinite life span is that I'm able to see how everything is recycled. Stars that die seed the next generation of stars with the heavy elements they produced in their cores. Galaxies that get ripped apart trigger the formation of more stars as they thrash in the fight for their lives. As long as particles can move around and interact, nothing in space can truly die. So, when I talk about the pain of feeling my stars die, what I'm really speaking about is the pain of guilt and failure. I'm sure you've felt that pain before, and you'll probably feel it again, but no failure of yours could ever possibly compare to mine, which have been recurring and billions of years in the making.

The first stars in the universe were composed of hydrogen and helium, the first elements created after the Big Bang, well before I was born. I got to see them, hold them within myself, even though I didn't create them. Or if I did, I don't remember it. To me, the stars were perfect. They were bright spots in the hot darkness, and I wanted to make some of my own. I didn't know yet that stars could die.

It wasn't obvious to me how to make a star, and the universe didn't provide any sort of instruction manual, so I had to improvise. I took stock of what I had, which wasn't much in the beginning—mostly just gas, a little bit of dust, and thankfully enough dark matter to hold myself together—and experimented.

Through trial and error, I figured out that if I squished enough gas into a small enough space, it started to shine. If I used too little gas, the temperature and pressure weren't high enough to trigger the fusion reaction in the core that supplies the energy that makes it glow. Since you've never actually witnessed fusion yourself, let alone initiated it, I suppose I should tell you how it works. Normally, explaining the minutiae with such underwhelming pupils would feel like a chore, but it's exciting to share this, the first thing I ever taught myself how to do, with someone else. Even a human.

Let's begin with the fact that your human scientists have identified four fundamental forces that they believe explain all the basic particle interactions in nature. Maybe there are more, but that's for me to know and your species to find out. Or not. The first force is gravity, which I would hope you're at least passingly familiar with since it's the force keeping you from being flung off your tiny rock as it rotates and moves through space. Gravity is the weakest of the forces by a wide margin, and it's also the only one that your scientists can't explain with a particle in their standard model of particle physics. Coincidence? Maybe. But also, maybe not.

The second force is electromagnetism, which dictates how charged particles interact with each other: like charges repel each other and opposite charges attract. Given your paltry human education, this is likely the extent of your knowledge of the four forces, and neither of the forces mentioned above are even responsible for nuclear fusion!

There's a third force that human scientists call the weak nuclear force, which is responsible for the radioactive decay of atoms. For example, the weak force can turn a neutron into a proton by changing the flavor of one of the particle's quarks. Don't even ask. I don't have time to explain quarks to you. They're too small, and talking about them makes my head hurt—or at least it would if I had a head—so you'll have to either learn about them yourself or hope that a proton chooses to write an autobiography, too.

The fourth force, called the strong nuclear force, is the one that's important for nuclear fusion. It's also responsible for holding the protons and neutrons in an atom's nucleus together. Though this is the strongest of the four forces by a loooong shot (6 duodecillion, or $6x10^{39}$ times stronger than gravity), it works on only the smallest of scales. Atoms must get extremely close to each other for the strong force to overcome the repulsion from electromagnetism, so close that it only happens naturally in the dense, hot cores of stars. It's almost like each atom is holding a tiny shepherd's hook that it can use to latch onto another atom, but only if they get close enough to lock the hooks together.

The fusion process could almost be called traumatic for the atoms involved. Their nuclei are first broken apart before they can be combined into a new atom that is heavier than both input atoms, but lighter than the sum of their masses. The missing mass gets converted to energy, and *that* is what makes my stars shine.

It took humans until the early 1900s to ascertain how nuclear fusion works. That's at least two hundred thousand years of trying to understand what makes stars glow. And rest assured that they *were* trying. It took me less time than that to *make* my first star, but making the star is just the first step.

(Well, actually, the real first step is cooling down enough gas so I can squish it into shape, but that's like saying the first step of cooking dinner is gathering the ingredients. It's so obvious that it shouldn't need to be written in the recipe.)

I learned that stars require a delicate balance between gravity pulling in and pressure pushing out. Your astronomers call this balance "hydrostatic equilibrium," a term, I'm convinced, they only would have chosen if they didn't want you to understand what's going on. The gravity comes, of course, from all the mass near the interior of the star, pulling in tandem on each individual gas particle closer to the edge. The pressure comes from the particles moving around inside the star, fueled by the burning fusion happening in the core. Minute changes in temperature, density, or fusion rate can knock the star out of balance and cause a cataclysmic reaction. In this way, I killed plenty of stars in those early eons, but they were acceptable casualties on the road to understanding. I made my peace with those losses.

I also concluded that it's much simpler to make stars in batches.[2] Maybe this is obvious to you as a human, since your brevity necessitates efficiency. It's why you invented the assembly line, after all. I had all the time in the universe to make these stars, so I wasn't worried about being quick, but I didn't want my new stars to be lonely. Instead of collapsing small gas clouds into individual stars, I started collapsing giant gas clouds to form clusters of around a thousand. Given enough time, most of these clusters will break apart as they orbit through my disk. These are your astronomers' open clusters, and they're distinct from globular clusters, which have more stars and are typically older. That's because many globular clusters weren't created by me at all but are instead the remaining cores of galaxies I devoured long ago. The gravitational attraction between stars in globular clusters holds them together over long timescales. I can't blame them. If another galaxy somehow managed to best me in battle, I'd hope the stars from my bulge would stick together, too.

After a few billion years' worth of stellar experiments, despite everything I had learned, the stars kept dying. Not all of them, but enough that I couldn't help but feel like it was all my fault, like I had made some fatal flaw in my creation. I went back to the metaphorical drawing board to design my next suite of experiments. (Galaxies don't have whiteboards or notebooks or spreadsheets to keep track of our thoughts like your scientists do. We're required to actually, you know, *remember* things.)

I went through a period of rapid star formation, creating stars more quickly than I ever had before in the hopes of finding the winning formula, the right combination of mass and metals that would make an undying star. I can only describe that period of my life as a manic episode. Human astronomers who study my past have noted that this burst happened around eight billion years ago. I also significantly slowed my rate of production right after that. In some other galaxies, this drop-off in star formation—a *quenching* process, as your human astronomers call it—happens when the galaxy has used up all of its gas and doesn't have a way to obtain any more. This is common in old elliptical galaxies that have already guzzled up the available gas. It's obvious even to human scientists that this is not what happened to me because I did eventually pick up my star formation rate about a billion years ago. (And it should be obvious to you that I didn't completely stop making stars in between the two bursts because your own sun is nearly five billion years old.)

Your astronomers have their guesses as to what quenched my star formation. Some of them think my disk was so warm from radiation coming off low-mass stars spread throughout it that gas clouds couldn't cool down enough to form new stars. Others maintain that my central bar—literally just a big block of stars whose orbits maintain a seemingly solid block of rotating mass—swept up all the gas and hoarded it so no stars could form (though they don't ascribe it so much agency in their explanations).

They're all wrong, of course. I simply stopped making stars because I was depressed. I *am* depressed, since that's not something that really goes away. It's something you learn to live with. And if you're a galaxy like me, you live with it for a very long time.

After conducting billions of experiments—varying mass, composition, and even location of the stars—I realized that every one of my stars was destined to die one way or another. Not only had I failed to make the immortal star I wanted, I had also apparently failed to identify a sound and reasonable scientific endeavor. My work was meaningless, and I needed time to wallow in that pain.

I know now—and your astronomers know it, too—that nearly everything about a star's death can be predicted if you know just one thing about it: its mass.

By my estimation, low-mass stars die the slowest and quietest deaths of all, much more of a whimper than a bang, and your astronomers agree. They can take trillions of years to fuse all their hydrogen into helium, so I've never actually seen one die, but I can do math and make an educated guess as well as the next galaxy. Better, even. These low-mass stars used to be my favorites. I thought they were triumphs, and now the knowledge that I'm just delaying the inevitable pain taunts me. But at least they'll buy me time at the end.

Precision is important here. By "low mass," I mean stars between 10 and 50 percent of the mass of your sun with temperatures between 2,500 K and 4,000 K. Human astronomers call them M-type stars or red dwarfs,[3] and they're the most common type of star in my body. Human astronomers who counted stars of different masses have noticed that more massive stars are less common. They call the distribution of stellar masses the initial mass function (or IMF), but they don't all agree on what the function is, or if there's a universally "right" one. If you ever want to cause an uproar among your astronomers, stand in a crowded planetarium and claim that the Kroupa IMF is better than the Salpeter. Most won't be able to refrain from loudly asserting their opinion back at you.[4]

I make more M dwarfs than any other kind of star because I know they'll last longer, even though I also know that they won't produce heavy elements or return much of their gas to be used later. They're selfish, but I like them.

To explain *why* red dwarfs don't produce heavy elements, we must look to another facet of my stellar experiments, one that focused on how to transfer heat. The majority of a star's energy is produced in its core, remember? But I wanted to build a star that would be hot and shiny all over, so I had to figure out how to move heat from one part of the star to another.

Heat can be transferred in three different ways:

- CONDUCTION, the transfer of heat through material in direct contact, like when you burn your delicate human hands on...well, anything even a little bit hot, really. You should have evolved to have thicker skin.
- CONVECTION, the transfer of heat through a fluid, like when you boil water because you need to...make food, I guess. I try not to think about things so disgustingly corporeal.
- RADIATION, which in this context means emitting electromagnetic waves that can carry energy through any medium, even the vacuum of space. Thermal radiation is why you sneaky humans can use infrared goggles to spy on people at night. A time when you should be paying attention to me.

I learned early on that conduction isn't that effective in stars. You see, stars are made of plasma and gas, which are both fluids (along with liquid, but stars aren't wet), and if you've ever boiled a pot of water, you know that convection is best for moving heat through things that flow. Radiation, a process where low-energy photons are generated to carry away heat through any medium, including space, works as well.

Red dwarfs are special because they transfer all their heat through convection. As hot blobs of plasma move away from the churning core, they find themselves surrounded by cooler material. The hot blob can then expand, become more buoyant, and float to the outer layers of the star even faster. At the same time, cooler blobs of gas at the surface contract and fall towards the core. This continuous cycle of motion mixes the material in the star and prevents helium from settling in the core, but it also transports hydrogen from the star's outer layers to its center. This innovation of mine made the M stars the most efficient hydrogen furnaces I'd ever created. Eventually, most of the hydrogen fuses into helium, but M dwarfs aren't massive enough to create the high-pressure conditions that trigger the next step in the nucleosynthesis process: fusing helium into carbon.[5]

Without fusion to generate outward pressure in its core, gravity wins and the red dwarf deflates in on itself. I did tell you the hydrostatic equilibrium was delicate. What's left behind is called a white dwarf, and it will slowly radiate away its heat until it's too cool to shine and no longer worthy of my notice.

One day, many billions of years from now, I won't have any gas left to make new stars. The massive ones will die, and I'll be left with just the dwarfs. I'm anticipating it will be a lonely time, but it's a long way off, even for me.

In the meantime, I'll continue to feel stars like your sun die their boring, anticlimactic deaths. These stars, which your astronomers *should* call G-type stars (that's the official classification, but you humans too often insist on calling them "sun-like" as if your sun is what all G-type stars strive to be), take about ten billion years to burn through all their hydrogen.

Unlike fully convective red dwarfs, G-type stars have a radiative core surrounded by a convective layer, which was tricky to build. Not only does each layer have its own average density, with the outer layers being less dense than the inner ones, but they also each have their own density gradients. The radiative layer of a G-type star is so dense and has such a steep gradient (meaning the change in density throughout the layer is extreme), convection isn't possible. Blobs moving away from the core find that they are much denser than their surroundings, so they fall back towards the center before they can reach the star's outer regions. Without convection, the only way to transfer heat through this layer is electromagnetic radiation. In other words, photons of light carry heat from the core to the outermost layer, which is diffuse enough for convection to happen. Unlike their lighter counterparts, G-type stars can produce the right conditions to make heavier elements like carbon and nitrogen.

When a G-type star fuses all the hydrogen in its core, it's initially too cool to burn helium. Lacking fusion's outward-pushing radiation pressure to maintain hydrostatic equilibrium, the star starts to contract, but as density increases, so does temperature. The star then quickly gets hot enough to fuse helium into beryllium ever so briefly before the new atom combines with another helium to make carbon. Then, like a drowning human given oxygen, the fusion makes the star start to expand. This happens several more times over the next billion years as the star burns through successively heavier elements. Human astronomers call these expanded G stars "red giants."

Your sun will puff up into the red giant phase in about 4.5 billion years. Humans will likely be long gone by then, wiped out by yet another of your planet's mass-extinction events. Believe it or not, that's a good thing, because if you stick around, you'll probably be engulfed by your sun's fiery envelope as it expands out into your solar system.[6] I'd say good luck, or have you wish on one of my "shooting stars"—which aren't even stars at all, just meteors—as seems to be your custom, but it wouldn't really do you any good.

After it (probably) destroys your entire planet, your sun, like other G-type stars, will use its stellar wind of charged particles to slough off its outermost layers. This is *very* cool to watch. The first time I witnessed it, I thought it was the star's grand finale. It wasn't. The core that gets left behind after the star sheds its bulky exterior collapses in on itself to form... *drumroll, please* a white dwarf!

The deaths of G-type stars may be disappointing, but the most massive stars die truly catastrophic deaths, and these failures are the hardest for me to bear. It's difficult for me to write about, too, but I'll do my best (which, for all intents and purposes, is also *the* best).

"Massive" is a peculiar word because it's relative. A star ten times the mass of your sun is certainly massive by your standards, but it wouldn't be if you lived around a star one hundred times the mass of your sun. That's totally hypothetical, of course, because these most massive stars barely survive long enough to form planets, let alone support life long enough for it to become—and I'll be generous here—intelligent. Just ten million years and then they're gone. They also give off ultraviolet and gamma rays that seem to do bad things to your fragile bodies, so you wouldn't last long if your species managed to develop there, anyway.

Human astronomers have avoided this relativity issue by abandoning nuance at the heaviest end of their stellar classification scheme. Just as radio waves are the big catchall category for all wavelengths longer than a millimeter, these so-called O-type stars, or blue giants, account for anything more than fifteen times the mass of your sun. They're plenty hot enough (at least 30,000 K) to ignite helium fusion in their cores even though they have an inner convection layer surrounded by an outer radiative layer, the opposite of intermediate-mass stars.

Maybe your astronomers lumped such a wide range of masses into a single category because high-mass stars are so rare? If so, that's probably my fault. Massive stars are hard to make, what with their unhappy endings. For me and the star.

The most massive stars I've made are sized between 150 and 200 solar masses. I've talked to other galaxies, and if they can make heavier stars, they're keeping it a secret. Since galaxies aren't prone to secrecy, owing to our long lives, and lies have a bad habit of building over time, I'm confident that stars can't get much bigger than that. Anyone who says they need something bigger is just trying to compensate for something.

There was that time a few billion years ago when Trin tried to convince everyone that it was *totally* possible to make a 300 solar mass star.

Whatever. We all know Trin is full of hot gas. It's hard to maintain hydrostatic equilibrium in anything heavier than 200 solar masses. That might seem counterintuitive to you, that gravity should lose out as a star gets more massive. Gravity is the weakest of the fundamental forces, remember? Radiation pressure, mostly from electromagnetic radiation in the form of photons, gets too strong at high masses. One of your human scientists figured this out without the help of hands-on experiments nearly a century ago, though he was more focused on the upper limit of a star's luminosity, or brightness, rather than mass, but the two are closely related. His name was Arthur Eddington, dubbed "Sir" even though he never fought in any battle on principle. Since I'm compelled to destruction, I must respect his steadfastness.

If you expect a 15 solar mass star to die the same way as a 100 solar mass star, then you clearly haven't been paying attention. Mass differences matter! It mattered when it was a difference between 0.5 and 1, and it matters when it's the difference between 15 and 100. Massive stars may not all die in the same way, but they do all make the same stop on their way to death, and that stop is a supernova explosion. More specifically, it's what your astronomers call a type II supernova, because you humans really do love to categorize things.

By the time an O-type star is ready to die, it will have fused all the hydrogen in its core into helium, which in turn was fused into carbon, then nitrogen, oxygen, silicon, and finally iron. These elements form in layers with iron at the core and hydrogen at the surface. Atoms bigger than iron are too big for the strong nuclear force to take over and aid in nuclear fusion, so heavier elements need to be fashioned in more cataclysmic events, like the collisions of neutron stars.[7]

When there's no more silicon to fuse into iron in the core, hydrostatic equilibrium breaks and the star collapses. This leaves too much matter packed into one dense space, so the star explodes. One might call it beautiful. Then one might remember that beauty breeds pain.

The explosion rockets all those heavy elements made by the O star out into the interstellar medium, the place between stars. Into me. I can use those heavy elements to create future metal-rich stars. In this way, the O-type star is the least selfish of them all.

As I said, the supernova was a stop on the way to death, not death itself. Low-mass O-types—see, it's all relative—leave behind dense remnants your astronomers call neutron stars. I suppose that's a reasonable name because these leftover cores are so dense that all their protons and electrons combine to form neutrons. They also make neutrinos, but do you really care what those are? I don't.[8] A neutron star, however, is as dense as if your sun were compressed to the size of one of your cities. Any city will do, except for Los Angeles, which is too much of a sprawl for my taste.

The heaviest stars leave behind even denser remnants: black holes. Objects so massive and so small that you can pull yourself away from one only if you're moving faster than the speed of light. They're as inescapable as death, taxes, and amateur stand-up comedy shows.

For thirteen billion years, I've been making stars and waiting for them to die, some more gloriously than others. Along the way, I've had to learn through my stellar experiments that lighter stars like M dwarfs live longer, but heavy ones like O stars give more back to the system.

They're all precious to me, though, regardless of their classification—you know, the M-, G-, and O-type system. This scheme that I've deigned to use was developed by a human named Annie Jump Cannon. Cannon divided the stars she observed into seven different categories based on their surface temperature.

From hottest to coolest (which also happens to be from most to least massive), they are O, B, A, F, G, K, and M. And she did it by looking at scribbly lines while working in a time—the turn of your twentieth century—when your species severely undervalued women's contributions. You could never!

These spectral types can be mapped onto a certain kind of plot that modern human astronomers learn about early in their careers (and then bring up again and again and again...). This plot, named the Hertzsprung-Russell diagram after the two scientists who separately conceived of it in some form or another, graphs the inherent brightness of stars against their stellar type as defined by Cannon's scheme. It's a surprisingly clever way of presenting the data so that a human can easily spot any patterns, because of course there are patterns.

Any human looking at this plot would be able to see that most stars fall on a curvy line that sweeps from the cool, dim M stars up to the hot and bright O-types. Your astronomers call this line the main sequence. Again, not a great designation, but we've already discussed your kind's naming limitations. This is just the group of stars that are actively fusing hydrogen into helium in their cores. Once they stop fusing hydrogen, they move off the main sequence and follow some interesting evolutionary paths. But every path ultimately leads to the same thing: failure.

You might be wondering why I kept trying to make new stars if I knew I was going to fail. Better yet, you might be wondering why I considered it failure if I knew it was inevitable that every star I made was going to die. Both questions share an answer, and it's a simple one, really: I love my stars.

I'm sure you probably think of me as a cold, unfeeling creature (which is at least a step up from not thinking of me at all), but that couldn't be further from the truth! I feel a confusing mix of despair and pride every time I best another galaxy in a fight. I feel gratitude for Sammy's company and frustration over Larry's presence. And I love Andromeda. Is it so hard to believe that I might love my stars, too, after spending billions of years trying to make every single one of them the best that it could possibly be? I should hope not.

But underneath it all, deeply central to my being and scattered throughout, is the guilt-ridden self-loathing that comes from knowing I'm condemning each of my stars to eventually die just so I can live. My whole body is peppered with bottomless pits of despair, figurative black holes that conveniently overlap the literal ones, both of them impossible to escape.

CHAPTER NINE

INNER TURMOIL

Most of my black holes are tiny, created when the most massive stars die their extraordinary deaths. These blemishes on my self-esteem each only hold the mass of maybe a few dozen Earth suns, but I have tens of millions of them scattered throughout my body. Any one of these disappointments would be totally manageable on its own, but the combined weight of them is crushing. Humans crack under the pressure of minor setbacks all the time. Crying over spilled milk is more than just an expression for so many of you, but what I've come to realize is that the milk isn't the real issue. It's the milk and the lost keys and the canceled date and the rejected job application and all the other digs at your happiness that pile on top of each other.

Just as the people in your life seldom see all the small things that drag you down into a dark mood, human astronomers struggle to see what they ironically call my stellar black holes. When they're alone, low-mass black holes don't have glowing hot accretion disks or bright jets demanding that everyone pay attention to them. Those flashy shows are reserved for the most energetic black holes, or active galactic nuclei (AGN), as your astronomers call them. Sometimes a stellar black hole will happen to pass between you and a shiny background source in just the right way for the black hole's gravity to bend the source's light towards you, but that's relatively rare. Instead, your astronomers have to study those of my stellar black holes that live in pairs. Black holes leech off anything around them, so when they have a companion, they steal its material and use it to build an accretion disk that shines with X-rays.

X-rays, by the way, are special—or maybe I should say especially frustrating—to your astronomers. You've been studying and using them on Earth since the mid-1800s, but it took you another century to observe X-rays from space because most of them can't pass through your planet's atmosphere. The wavelength of X-ray light is even smaller than the molecules you breathe in, so X-ray photons can't travel far in air before they get absorbed. This means your successful X-ray missions must fly high above Earth's surface on high-altitude balloons or get launched into orbit, like NASA's Chandra observatory.

I digress. The point of this chapter isn't to talk about the small shames, but to tell you about the Big One, the so-called supermassive black hole at my center that your astronomers have named Sagittarius A*. There was a long period in my life, many billions of years, when it would have been nearly impossible for me to talk about this. I'm not quite to the point of acceptance, but it no longer hurts quite so much to think about my black holes.

So, I'll soldier through this chapter because my story deserves to be told. And you...well, you have so much more to learn.

I call my central black hole Sarge. I realized long ago that it was easier to confront something if I gave it a name. The name itself is a transliteration of the concept derived from the Old Galactic for...well, I guess the closest you have for it on Earth is "doody head." (It sounds less childish before the translation.) Remember, this was long before your astronomers even knew they lived in a galaxy, let alone that they shared it with a hateful, churning mass that lurks in the shadows of my guilt, waiting to corrupt and swallow anything that gets too close. Your astronomers just named it after the patch of sky where they detected Sarge's signal, so any similarity in the name is merely coincidence.

It's actually a funny little human story that started with this man named Karl Jansky. Nowadays, Karl is considered the "father of radio astronomy" and he even has his own unit named after him. The jansky measures flux density, or the amount of energy passing through a specific area in a given amount of time and then normalized by the bandwidth of the telescope receiver. There are other units of flux density used in other human sciences, but the jansky is a *special* unit that's useful only for exceptionally dim and small sources with broad continuum emission. It's basically used solely by radio astronomers, and there are only about one thousand of them. Plus the people who do use the jansky always seem to be complaining about it, so...that's not great for Karl.

In the 1930s—a period that seemed, er, rough for creatures like you who rely so heavily on eating but can stomach only a tiny fraction of what surrounds them—Karl found a radio signal coming from the region of space dubbed Sagittarius, one of eighty-eight such official constellations according to your astronomers' precious IAU.

Remember, your kind could just barely see into space back then. They had no idea how to see through the 8 kpc worth of dust, gas, stars, and black holes and everything else that can block, warp, or redirect the light that your astronomers study. So it should come as no surprise when I tell you it took them another forty years to find Sarge, as if I haven't been complaining about it for millions of years to anyone who would listen.

In those forty years, various astronomers figured out that Karl's radio signal was indeed coming from the center of their home galaxy—aka *moi*—and that it was in fact multiple overlapping radio sources, including one object brighter and denser than all the others. By your 1980s, human astronomers had collected enough information about Sarge to determine that it was likely a black hole, because they simply didn't know about anything else that could be so small yet so massive. By then, they were calling it Sagittarius A* because it was an "exciting" object in this central, radio-loud region, and your physicists had been using asterisks to denote excited atomic states. If they knew Sarge the way I do, "exciting" would have been the last description to come to mind. Then again, if any human *really* knew Sarge as intimately as I do, they'd have been ripped apart immediately, and I know some galaxies who would find that very exciting indeed.

Just to make it abundantly, human-proofly clear: Sagittarius is a region of the sky, Sagittarius A is a complex and multifarious source of radio emission found in the Sgr region, and Sagittarius A* is the brightest of Sgr A's parts.

Each of these discoveries had to be made one at a time, building on what the previous generation of human scientists had achieved. You humans move so slowly, devoting your entire lives to solving a single tiny part of just one of the many problems that plague your people. I love to see humans doing cute human shit like this.

But less about humans and more about Sarge. I'll be honest—when am I not?—I was dreading getting to this, so I had to ease my way into it.

Recall your oldest, sharpest, most shameful memory. The one with jagged edges that still elicits the slow creep of hot mortification up your neck every time it rips through your mind. It's probably something like that time you single-handedly lost the most important sports game of the season, or when you went on vacation for a week without asking someone to feed your pet gerbil. Whatever it is, thinking about it probably makes you uncomfortable. But it's just that: a thought, a fading memory of something that happened in the past.

Sarge, however, is the physical embodiment of *everything* I have ever hated about myself. Every galaxy eaten, every misguided flirtatious blunder, every undeserved snide comment directed at Sammy. (None of the petty remarks directed at Trin or Larry, though, because those were absolutely deserved.) And when I absorb other galaxies, I take on the weight of their shame, too.

The fact that we galaxies hold literal manifestations of our bad feelings within ourselves is, perhaps, the worst part of our lived experiences. I guess it's the flaw in our design that refutes the possibility of an intelligent creator. But if it's a choice between having a black hole and having to—*gag*—wipe a butt, I'm choosing the first hole every time. The other apes on your planet don't seem to wipe their butts, but you lot gave that up just so you could walk upright.[1]

Sarge is where my darkest, heaviest memories live, the ones that have sunk to the core of my being and do their best to drag me down with them. Every time I try something new—like the first time I made a planet with a ring[2] around it—Sarge is there to fill me with doubt. "A ring?" it asked. "What, are you too lazy to make a moon?" Whenever I send a star note to the Leos—I have a kind of group chat going with a family of galaxies who live around the block—Sarge is there to make sure I remember every grammatical error, every perceived overshare. "Are you sure they're still your friends after you put so much magnesium in that note? The spectrum is almost indecipherable!" Even now, it's telling me I'm too whiny and that no one cares about my life and my problems. No one important, anyway. I'm certain you're absolutely enthralled.

Sarge stops me from envisioning what I could be because I'm too busy lamenting what I've never been and despairing what I was instead. Of course, that's not how your astronomers think of black holes. To them, Sarge is just an intellectual curiosity, a dense astronomical mystery whose solution could earn them a fancy prize. In fact, three human astronomers were jointly awarded the 2020 Nobel Prize in Physics for confirming that Sarge is indeed a black hole. That certainly took them long enough! It's no wonder that human astronomers talk about black holes as if they're marvelous phenomena to behold; they haven't known Sarge long enough to understand its true nature.

Physically speaking, black holes are made of two or three parts. There's the black hole itself, which is the dense part in the middle

that light can't escape, and its outermost boundary is called the event horizon. Then there's the accretion disk, a ring of material that's slowly spiraling its way in towards the black hole and glowing because of all the friction between the individual particles in the disk. Finally, some black holes have what your astronomers call jets, which are bright and powerful streams of material shooting up and down—though such simple directions are typically meaningless in space—from the plane of the disk.

Your scientists use three different values to describe black holes: mass, charge, and spin. We've come far enough together that I trust I can skip an explanation of what mass is...

Charge refers to its electric charge, which is basically just the difference between the number of protons and electrons, or positive and negative charges. Black holes tend to be neutrally charged, since there are about as many protons as there are electrons in the universe, and they're both equally likely to be absorbed into the black hole and cancel each other out. In fact, your astronomers usually just assume that the black holes they study have no charge to make their calculations easier, even though the charge is constantly in flux as the black hole accretes—or eats—new material.

A black hole's spin, or angular momentum, is... exactly what it sounds like. Your astronomers occasionally do come up with sensible names! The greater a black hole's spin, the more it warps and drags space-time around it. Small stellar-mass black holes spin because they form from the collapse of spinning massive stars. (The stars in turn get their spin from the rotation of the gas cloud I used to build them, and the gas clouds... well, it's just rotation all the way down. Or up. Or out? See? Useless! Let's just say that pretty much everything in space is spinning.) Heavier black holes like Sarge get their spin from the momentum left over after the collisions necessary to build something so massive.

Sarge is just about four million times the mass of your sun, though human measurements of its mass have ranged between 3 and 5 million solar masses. Just like you repress your least favorite memories by confining them into the darkest recesses of your mind, I've squished that angry mass into a space smaller than your planet's tiny orbit around your sun. Some human astronomers say that it's even smaller than Mercury's orbit.

For those of you who need a little help putting the pieces together (which will probably be most of your kind, so don't feel too worthless), objects that massive and that small are too dense to let even light escape their gravitational pull. The "fabric of space-time," as your scientists are so fond of describing it, bends around that much mass and warps so that anything trying to escape, like photons or self-acceptance, gets turned around to face where it came from.

The most salient consequence of a black hole's extreme density, at least for humans, is that it's impossible to *see* them. Not that I owe you any explanation, but I swear I didn't do this on purpose. If I could control anything about Sarge, I absolutely would, if only to wrest back a modicum of dignity. Besides, even if I could have made Sarge visible to the human eye, it probably wouldn't have crossed my mind as something worth doing. No eyes, remember?

The fact that humans can't see black holes is how they got their name. The term seems to have leaked out into the human vernacular as something that sucks all the energy and life out of a room, which is accurate, but still leads to the common misconception among your kind that black holes are like vacuums, sucking all the material around them. False! A black hole would never put that much effort into anything. No, they're just pits for slow stuff to fall into. If your sun suddenly became a black hole with the same mass tomorrow, everyone you know would probably die shortly thereafter, but not because you got suctioned into the middle of your solar system. You'd continue on your same orbit until your planet and everything on it froze without the heat of the star I made that you're totally taking advantage of.

The term black hole also gives too many humans the foolish idea that there's a connection between black holes, dark matter, and dark energy, even though they are quite different. Black holes are extremely dense objects made of regular matter just like you and the shiny parts of me. They're the type of matter your scientists call "baryonic." Dark matter is...well, your scientists aren't totally sure what it's made of, but it behaves just like baryonic matter in every way *except* for the fact that it doesn't interact with light. (Some of your astronomers and physicists think that dark matter might be made up of little black holes, but it's not a very popular idea.) And dark energy isn't matter at all, but instead the name your scientists gave to the unseen force propelling the expansion of the universe. All invisible to the human eye, but then again, most things are invisible to the human eye, so that's not a good reason to lump them together.

"But, Milky Way," I hope you're saying, "how do we study the black hole if we can't see it?"

Remarkably astute of you to ask! Well, first, you need to accept that there are other ways of knowing about something without seeing it. Even if I weren't already always keenly aware of Sarge's presence, I would still be able to *feel* it sitting there, tugging on my stars and spewing toxic energy. Second, to actually answer "your" question, human astronomers study black holes by measuring the way they affect their environments.

Since the 1990s, your astronomers have studied infrared and radio signals (the wavelengths of light that can most easily pass through the dust between you and Sarge) to measure the positions and velocities of some of the stars orbiting closest to my center. Even human scientists understand that gravity drives motion in space and that gravity comes from mass, so learning about the motion of these so-called S stars can help your astronomers constrain Sarge's mass. I pride myself on creating exceptional stars, especially the ones brave and resilient enough to live that close to Sarge, but sadly your astronomers seem to want them only for the information they can provide. Rude, but I suppose they don't know any better. They've found one star, dubbed S2 because it's the second closest star to Sarge (that you know of!), to be particularly elucidating. S2 orbits Sarge once every sixteen or so Earth years, often enough that your astronomers have seen more than one full period of S2's very elliptical orbit.[3] These stars are close enough to Sarge that it makes the most sense to use your human unit AU. S2 spends much of its time at approximately 950 AU from Sarge, but at its most bold, S2 dips down to just 120 AU from the massive monster. At its closest, S2 must move at 7,700 kilometers per second, which is 2.5 percent the speed of light! It's the Usain Bolt, the speeding bullet of stars. My favorite little daredevil <3

There was an astronomer a few Earth centuries ago called

Kepler who spent a lot of time thinking about orbits of moons, planets, stars...they all work the same way under most conditions. (Those chaotic stars in my bulge tend to avoid moving on what your astronomers would consider Keplerian orbits.) Kepler figured out that if you know the distance and period of an orbiting body, you could calculate the mass of the object it orbits. Or the combined mass of the object*s* it orbits. Your astronomers used Kepler's work and S2's orbit to "weigh" Sarge.

It should come as no surprise to you that a black hole's mass determines its size. By definition, a black hole is an object so massive and dense that light can't escape, so a black hole of a certain mass can get only so big before it falls below the density threshold. Under ideal, simple conditions where the black hole has neither charge nor spin, that size cutoff is the Schwarzschild radius, or the distance from an object where the escape velocity equals the speed of light. Sarge's Schwarzschild radius is about one-tenth of an AU, but its actual radius is smaller than that.

Recently—and I mean recently for humans here, so it hasn't been long at all—your astronomers finally figured out how to take a picture of a black hole. Well, of the event horizon at the edge of a black hole. The signal they detected is called synchrotron radiation, produced by electrons accelerating around magnetic field lines, almost like they're screaming as they fall. To take the picture, astronomers had to build a telescope as big as your planet. The bigger a telescope,[4] the better it can see small objects clearly. And though even the smallest black hole would absolutely dwarf you in size, they look small to you because they're so far away. I wouldn't have exhausted that much effort just to take a picture of something so tinged by regret, but it must be some earthly rite of passage, since so many of you capture photos of your immediately lamentable awkward preteen years.

The smallest black hole your astronomers have ever found is just about three times the mass of your sun.[5] THREE! That's less than one-millionth of Sarge's mass, and only fifteen miles across. Understanding this lower mass limit for black holes will help human astronomers distinguish between black holes and neutron stars, the remnants of stars just slightly less massive than the ones that form black holes. The most important difference is density—neutron stars are dense enough that electrons are forced inside protons to produce neutrons, but your astronomers just aren't yet sure where that threshold is.

Back to this Earth-sized telescope. Obviously, your species isn't capable of building a single structure as big as your planet, though I would love to see the chaos that would ensue if you tried. Instead, your astronomers use powerful computers to analyze data from carefully timed observations taken by telescopes across your globe, similar to MeerKAT, but on a larger scale. In fact, the basic concept is over a century old—developed in the late 1800s—but had never been applied to a project of this magnitude. Until the Event Horizon Telescope.

The EHT combines telescopes from at least eight different Earth-based observatories, and they're adding more to the network over time. In 2019, after combing through literal piles of data[6] from the worldwide array of telescopes, your astronomers released the first-ever human-made image of a black hole's event horizon. But it wasn't Sarge's event horizon. It wasn't even the event horizon of any of my smaller black holes. It was some *other galaxy's* black hole. M87.

M87 is an elliptical living in the next galaxy cluster over, the Virgo Cluster of galaxies. (NOT to be confused with the Virgo Supercluster. Think of it as the difference between New York City and New York State, because galaxies in the Virgo Cluster are as proud to live in their specific metropolis as the brashest New Yorker.) As the biggest and strongest Virgo Cluster galaxy, M87 has a lot of responsibility, a long history of unsavory deeds, and an understandably gargantuan central black hole. Even though Sarge is closer to you, M87's black hole is easier for your nearsighted human instruments to image.

I'm not sure M87 appreciated being photographed like that without permission, but I also doubt your astronomers care much about getting galactic consent. That image held information about the size of M87's central black hole (and therefore its mass) and the direction of its spin (the side moving *towards* you will appear brighter because of the Doppler effect).

More observations revealed the strong, spiraling magnetic field lines around the black hole's event horizon. This supported an early hypothesis from the 1970s about the formation of jets, which were first observed in 1918 by Heber Curtis of Great Debate fame. Roger Blandford and Roman Znajek were working at Cambridge University when they speculated—without a shred of evidence and probably over a spot of tea—that spinning black holes can twist their magnetic field lines into a spiral. Voltage traveling along the lines draws energy away from the accretion disk, and the black hole gets to put on a light show.

M87 and I aren't the only galaxies with supermassive black holes. Every galaxy has one at its center. Well, all the *real* ones. Most dwarf galaxies don't have them, which makes sense. What would a galaxy that small have to be so upset about?

Still, there are some dwarf galaxies that do carry that weight around with them. Your astronomers have found a dozen or so, and they've run simulations on their computers—they do love those computers—to convince themselves that the black holes are real. They were especially interested in the supermassive black holes they found away from the centers of those guilt-laden dwarfs, since black holes usually try to worm their way into the middle of everything.[7] Their black holes are typically smaller, only about a million times more massive than your sun, but again, that's to be expected for a dwarf.

According to your astronomers' computer simulations, these off-center black holes should be fairly common for dwarfs who don't have the mass, and therefore the gravitational strength, required to hold them in place. About half of the supermassive black holes in dwarf galaxies have wandered away from the central point. But, surprise surprise, the simulations don't capture the whole picture.

Dwarf galaxy collisions aren't as violent as the ones I've experienced in the past. It's been quite some time since I went up against a galaxy who could put up much of a fight. But dwarfs, being the most common type of galaxy in the universe, encounter each other often. And when one dwarf bests another, it's a fairer fight, one that each side can be proud of. Usually. Some dwarfs are more prone to guilt than others, and sometimes the fighting gets a little dirty. When that happens, the embarrassment isn't enough to become the dwarf's core trait, the one that everything else revolves around. Hence the black holes exist away from the dwarf's center.

Dwarf galaxies' holes notwithstanding, as supermassive black holes go, Sarge isn't even that impressive. There are plenty of good, hardworking galaxies with bigger black holes than mine. M87's black hole is around six billion times more massive than your sun. A galaxy your astronomers call Holmberg 15A (named after the astronomer who first discovered it in 1937) who lives a few clusters over in Abell 85 has a central black hole roughly forty billion times your sun's mass, built up over many, many mergers with smaller, weaker galaxies. The heaviest black hole your astronomers have discovered is seventy billion times more massive than your sun, making it more than fifteen thousand times more massive than Sarge. This oversized monster can be found in a galaxy more than ten billion light-years away from us. It's so distant that your astronomers haven't even named the galaxy, just the quasar[8] at its center (TON 618), with bright jets powered by the gargantuan black hole. I've never met this galaxy, but even I shudder at the thought of what it must have done to build up such a foreboding black hole in so little time. (For those of you who still need me to spell it out, light travels at a finite speed, so you're seeing the galaxy as it was ten billion years ago, making its supermassive black hole impressively young.)

Sarge may not be the biggest black hole in the universe, but there have still been times when I thought it would devour me whole. Humans often talk about how their grief "eats away" at them, but for galaxies, the eating isn't metaphorical.

Sarge's immense gravitational pull guzzles down several

sextillion metric tons of material every year. That's ten Earths a year! That may not seem like much—or maybe to you it seems like too much—but I have lived for billions of years, and that little bit builds up over time. The material it swallows—the gas, the dust, even the outer layers shed by some of my stars—doesn't just disappear into the black hole. No, it is ripped apart and twisted in on itself until I wouldn't recognize it.

Your astronomers call this stretching and ripping "spaghettification." Of course they would need to give it a cutesy little name like that, but I can't imagine a stomach full of spaghetti feels like a barbaric torture scene happening in your gut. Black holes don't just have extreme gravity; they also have extremely steep gravitational gradients. In other words, the strength of the gravitational pull changes quickly as you move away from the black hole. When objects get close enough to a black hole, they experience the very real consequences of that gradient. Gravity pulls much harder on the side of the object closest to the black hole than it does on the far side. Even for something as small as you, the difference in gravitational force between your head and your feet is stronger than the force holding you together. You and anything else that ventures too close to one of these monsters would get stretched, spaghettified, right before you disappeared forever.

Galaxies like me need gas to survive. It's how we make our stars, and when we run out of it, that's the beginning of the end. That's why galaxies are always eating each other, because there's only so much gas in the universe, and most of it is caught up in our stars by now. Black holes like Sarge can suck up a galaxy's gas or use its gravity and feedback winds to fling the gas out of reach.

Worse than that, if left unchecked, a supermassive black hole can gorge itself on so much material and build up its accretion disk to the point where it starts to *heat up* the gas around it, making it much more difficult for us galaxies to form stars.

That's what happened to JO201 (henceforth Jo), a large spiral galaxy living in Abell 85, making it one of Holmberg 15A's neighbors. Sadly, Jo was overwhelmed by the crushing weight and negative energy from the supermassive black hole at its center. The black hole stole or heated up so much of Jo's gas that it couldn't make any more stars. It would have been so easy for Jo to do nothing and let its black hole finish the job. But about a billion years ago, in a last-ditch effort to overcome its black hole's death grip and kick-start some star formation, Jo started careening towards the center of Abell 85 faster than the speed of sound. "Supersonic," your scientists call it. Abell 85 is a large galaxy cluster, home to about five hundred galaxies. Jo knew that passing through a dense environment like that (not as dense as a black hole, of course, but still packed together compared to the pressureless vacuum of space) at such high speeds would force the gas near its edges to mix and make new stars. This dive is only a temporary solution, especially if Jo's black hole continues to grow unimpeded. But Jo's a tough, resourceful galaxy, so I'm sure it will figure out how to coexist with its black hole.

That same pressure—called "ram pressure" for some reason—that forces the gas to mix also causes big tendrils of gas to flow behind Jo like a cape, like the hero it is, as it makes this lifesaving journey. Your scientists call Jo and other galaxies with these dragging tails "jellyfish galaxies." I've studied Earth's jellyfish, and I

admit there's a physical resemblance, but I doubt that any of your marine jellies, with even less ability to comprehend existence than you humans, would be able to empathize with what Jo's going through right now.

What's happening isn't Jo's fault, though not all galaxies are gracious enough to see it that way. Your astronomers say that a galaxy losing its gas is quenching. I call it suffocating or starving, maybe even choking. Whatever you call it, it's a slow and painful process. It's a death that can announce itself far in advance because most galaxies don't know how to stop it even if they see it coming.

I, of course, am not like most galaxies. I realized that if I can't control anything Sarge does, I can control everything I do *around* Sarge. Just as you can't control the world around you, but you can control your response to it. Except...reversed. Either way, I can't make Sarge less massive or slow its rotation, but I can try to direct my stars and gas away from it to limit its growth. And I can slow their orbits so that they don't add to Sarge's angular momentum.

For me, the question was never whether I could beat Sarge, but if I wanted to.[i]

---

i This is a note from Moiya. If you, like the Milky Way, are doubting your desire to stay alive, know this: You are not alone, and you would be missed. Please seek help. I know it's hard, but it's worth it. YOU are worth it. <3

CHAPTER TEN

AFTERLIFE

Perhaps, if I weren't so burdened with the heavy weight of knowledge and the curse of all kinds of brilliance, I'd be able to do what Jo didn't and convince myself that dying would take me to a "better place." But alas, the idea of a happy afterlife is too human a comfort. It's a solace that relies on faith, which is ultimately the admission that there are forces at play beyond one's understanding. And while there are some things that I do not *know*, I have yet to come across anything in my thirteen billion years of living that I cannot *understand*.

But in the quiet millennia when work was in a lull, Sammy and everyone else were minding their own business, and the only voice I could hear was Sarge's taunting whisper, I used to dream about a universe with different rules. I imagined a different kind of life with less grief and responsibility, more dancing and an endless, violence-free supply of cool gas to munch on. Maybe a kind of next stage of existence where I shed the worst parts of this one but remember all the lessons I've learned along the way. Similarly, it seems you humans have appraised the many, many harsh inconveniences of your reality and imagined your own next step... in a variety of directions.

Hundreds of thousands of years ago, when the human branch of Earth's evolutionary tree still had several surviving offshoots,[1] early human species buried their dead in pits or disposed of them in deep caves or sailed them out to sea. Your modern archaeologists and anthropologists and others who try to glean ancient intentions from fossil evidence aren't sure if those practices belied a belief in life after death. It's possible that early humans understood the dangers of leaving a corpse exposed where dangerous animals might come looking for it.

Over time, your burial practices became more elaborate.[2] You started saying specific words over the dead, preparing their bodies in ritualistic ways, and marking their graves so that you could return to a place that was no longer a mere disposal site, but an eternal resting place. You buried the bodies with both practical and precious items like food, clothes, and gems, and you formalized grieving rituals. Your archaeologists aren't sure when your ancestors started believing in another life after death, but they've found evidence of burial offerings going back at least one hundred thousand years. Even those humans understood that death was a one-way exit from this world, so why else would they bury these useful objects with the dead if not for a belief that there was *something* on the other side of that door?

Humans' enduring belief in the afterlife has served several purposes. It helped you rationalize death before you understood the science behind your delicate transience. It comforted humans whose loved ones had died and who knew they would eventually meet the same fate. If I lived a life as short as yours, I, too, would love to think I had a chance to be reunited with my favorite galaxies and stars. Eventually, the concept of life after death also provided many religious and political leaders with an effective way to enforce societal norms. "Be good, or your immortal soul will suffer for eternity in hell." Of course, not all ancient humans believed in an immortal soul—that kind of abstract thought wasn't even possible on macro scales until your brains grew space for it some fifty thousand years ago—and only some of them believed in something like a hell, and only a fraction of *them* believed that the soul would stay in hell forever. The point I'm making here, if you would stop getting distracted by details, is that the threat of punishment after death—or conversely, the promise of a reward—kept humans in line while they were alive. Being that I'm responsible for billions of stars, I must say I understand the impulse, and might have even picked up a managerial trick or two.

But the most important purpose that your afterlife myths have served is, obviously, entertaining me. In a popular one, the new glorified bodies you get once you reach heaven are imperishable and powerful. I've always wondered what kind of shenanigans your creative human minds could cook up if you didn't have to worry about keeping yourselves intact. And since early humans in their scattered tribes each developed their own rendition of the world beyond death's door, there are so many afterlives to choose from!

Your ancient Egyptians told many tales about what to expect

after their earthly bodies died, and even wrote instruction manuals so that the body and soul of the deceased could be properly prepared for the journey. Maybe you've heard of it, the Book of the Dead, though in reality it wasn't a standardized book. Many families had their own, like yours might have its own cookbook. Instead of detailing recipes for mac and cheese, their books described how to prepare a person's soul and body for the next stage. The soul was guided to the afterlife with prayer, and the body was preserved through mummification, the painstaking process Egyptians developed to ward off decomposition—such a disgustingly... organic phenomenon. The bodies needed to be kept in pristine condition, for it was written that the soul would need a body to inhabit in the next life.

Once in the so-called underworld, the dead would have to confess their sins to a panel of judges and have their hearts weighed by Osiris against a legendary feather of justice. If they failed these tests, their hearts were eaten by a chimeric beast with the head of a crocodile, and their souls ceased to exist. But if they passed, they could spend the rest of their immortal existence sailing across the sky, traveling through my glorious celestial form with the sun god, Ra; or they could stay in the underworld with Osiris; or, as was most common in the stories, they could stay in the Field of Reeds (some modern humans have translated this as the Field of Rushes), where they would live a life much like the one they led on Earth, but with their own land and possibly many servants. I know which place I would choose to spend eternity, but I'm not here to fault you for your simple human pleasures.

The Greeks talked of Hades, Hindus told stories about their

souls being reincarnated into new earthly bodies, and the Norse looked forward to dying as warriors so they could drink and train for battle in Valhalla. None of these prospective fates moved me like those from cultures that placed their afterlives among the stars.

Few human groups knew how to catch and hold my attention like the Maya. They didn't just tell stories about me; they made me a part of almost every aspect of their lives, from planning their cities so that sacred structures like temples and palaces would line up with objects in the sky, to venerating astronomers for their knowledge of celestial movements. But their myths were the most touching homages to me, and one particular group of Mayan peoples, the K'iche', told tales that featured me as the road to the afterlife.

They spoke of Xibalba, an underground place full of demons and dangerous trials. Humans could hope to reach Xibalba alive by sneaking through a special cave, but the bravest and most powerful K'iche' traveled to Xibalba like their sun god, Kinich Ahau, who turned himself into a jaguar every night to stalk through the underworld. And how did he get there? Through me, of course!

Your ancestors have so heavily polluted your sky with light and smog that some of my best features are hidden to you, but the K'iche' in the ninth century saw a dark path that cut through the luminous stream I cast across your night sky. They called that path the Road to Xibalba.

The K'iche' told one story that has stayed with me these past thousand years, a story about a pair of twins named Hunahpu and

Xbalanque. The twins were invited by the petty lords of Xibalba to play ball[3] in their underground court, where they were met with one deceitful trick after another. The lords forced the twins to play with a ball studded with razor-sharp spikes, locked them in a dark house filled with knives that moved on their own, and even managed to cut off Hunahpu's head. But the twins devised an impressively complex plan to defeat the lords of Xibalba. In this scheme, they allowed themselves to be killed, were reincarnated as young boys, and performed miraculous acts until they were reinvited to perform for the Xibalban lords in their new, unrecognizable forms. They then used the element of surprise to kill the lords of death and set free the K'iche' people from lives of forced servitude to the evil demons. In some versions of the story, Hunahpu and Xbalanque become the sun and moon.

The K'iche' call the twins heroes, and there are similar stories among other Mayan groups. But I think we all know who the real hero of this story is. After all, Hunahpu and Xbalanque wouldn't have made it to Xibalba in the first place without me there to show them the way.

They differ so widely, but the one thing that nearly all your afterlife myths have in common is the way they condemn people who step through to the next stage of their own accord. It is logical that these stories would discourage *any* destruction of human life. But as I said, this afterlife stuff is human folly. The threat of fire and brimstone isn't as compelling to a galaxy like me who makes giant fusion factories for a living, so I had to convince myself to stay alive the old-fashioned way.

After billions of years of holding on to myself for dear life lest

I succumb to Sarge's taunts and spiral down into my despair past the point of no return, I grew tired of the struggle. I had watched too many friends buckle under the pressure from their black holes. More than that, I had lost too much of *myself* and wasted too much of my precious time believing Sarge's twisted lies.

I remembered that I'm the Milky Way, damn it! The greatest galaxy in the Local Group, except for one very special spiral. Sure, I've done some terrible things, but I did them to survive. I've failed often—I'm sure I'll fail again, plenty of times—but at least that means I tried. And after watching my most selfless stars live out their too-short lives, I realized that I didn't want to pass up the opportunity to live for a trillion more years. Otherwise, they've died in vain.

Those thoughts usually give me the strength I need to hold Sarge at bay.

Sometimes I slip up and forget that Sarge doesn't define who I am as a galaxy. The black hole's activity flares[4] and the doubt creeps back in. When that happens—when I can't block the thoughts out for my own well-being—one thing keeps me going. I remember that there's another galaxy out there reaching out, one struggling to defeat its black hole and calling for my help. Okay, there are billions of galaxies like that, but there's one I care about orders of magnitude more than the rest: Andromeda.

CHAPTER ELEVEN

CONSTELLATIONS

HUMANS HAVE KNOWN ABOUT ANDROMEDA since you first bothered to look out at the rest of the universe. The Persian—or is it Iranian? Your imaginary, vacillating borders change too frequently for me to keep track—astronomer Abd al-Rahman al-Sufi wrote about Andromeda as one of several nebulous smears in his *Book of Fixed Stars* in the 900s. As if the most perfect assembly of stars to ever grace this universe could be compared to a mere globular cluster! I'm glad to see that your modern astronomers have much more respect for Andromeda, whom they know by many names: Messier 31 (or M31), NGC 224, IRAS 00400+4059, 2MASX J00424433+4116074... They aren't the most appropriately poetic names for a galaxy as magnificent as Andromeda, but these monikers aren't supposed to rouse emotion. They're meant to encode information about the galaxy's position and the telescope or survey that observed it. The name Andromeda, however, comes from an ancient Greek myth about an Ethiopian[1] princess, which

I'm sure you'll assume is some sort of compliment. We'll see what you think after you hear the story.

Princess Andromeda was born to two parents (as is customary for your species): King Cepheus and Queen Cassiopeia of Ethiopia. Andromeda is a Greek name meaning "ruler of men," so I doubt it was truly the name of a real Ethiopian princess, but we'll ignore that oversight, along with all the other nonsensical parts of mythical stories, and chalk it up to creative license.

Given the name they picked, it should come as no surprise that Cepheus and Cassiopeia were proud people with high expectations for their daughter. In fact, Queen Cassiopeia boasted to anyone who would listen that Andromeda was more beautiful than the Nereids, sea nymphs renowned for their good looks. Gasp!

As the story goes, Poseidon, the god of the sea, was so offended by Cassiopeia's hubris that he flooded Ethiopia's coastline and sent a sea monster to terrorize the kingdom. I would have thought that a god responsible for all of your planet's seas and seismic activity would have more important things to do than punish someone for a minor slight based on a wholly subjective scale of aesthetic pleasure, but such are the contrivances of mortal creatures trying to imagine the inner machinations of immortal gods. King Cepheus trekked through the desert to consult the oracle of Ammon about how to rid his kingdom of the destructive monster, Cetus. The oracle told Cepheus that the only way to stop the carnage was to offer Andromeda as sacrifice to the monster. And Cepheus, in a move that defied every parenting instinct I had come to expect from your kind, *agreed*! He went back to Ethiopia and chained his daughter to a rock by the sea. I would never treat my stars with such callousness, and I have many more to spare.

Don't worry, that's not where Princess Andromeda's story ends, no thanks to her. The hero Perseus, flying high in Hermes's swift-winged shoes after beheading the Gorgon Medusa (yet another woman wronged by Poseidon[2]), happened upon Andromeda on her little rock and fell madly in love. Is that really all it takes for you mammals?

Apparently it is, because Perseus made a deal with King Cepheus to marry Andromeda if he could slay Cetus. He did, of course, because he wouldn't be a worthwhile hero if he could be bested by your run-of-the-mill sea monster, and won the princess's hand in marriage.

Andromeda and Perseus spent the rest of their lives as monarchs and had a chaos of children and many impressive descendants, including Hercules (or Heracles), whom I'm sure you must have heard of. When Princess-turned-Queen Andromeda died, the goddess Athena put her in the sky as the Andromeda constellation.

If you knew Andromeda like I do, you would understand that similarities between the princess and the galaxy are in short supply. Both are exceptionally beautiful—I don't find the human form particularly pleasing, so I'll have to take Cassiopeia's word for it—but the comparison ends there. Where Andromeda the princess is weak and passive in deciding her fate, Andromeda the galaxy has no qualms about exerting its will on whatever is within reach.

It's a good thing that Andromeda the galaxy wasn't *technically* named after the princess, but instead after the constellation, which was named after the asterism, which, according to the story, was the celestial embodiment of the princess's spirit, so...we've come full circle and that means we should move on.

If you want to see the Andromeda constellation—and the smokeshow of a galaxy enclosed within—you'll need to be on the right part of your planet, which shouldn't be too difficult even though so much of Earth's surface is covered in water that you never learned to stand on. (Though I've heard some rumors about at least one human walking on it...) If you're positioned at a latitude of roughly 40° north between the months of August and February, Andromeda will pass right over your head in the night. Or late evening / early morning, depending on the date. To find Andromeda, you can hop from the Cassiopeia asterism or from the star your astronomers named Alpheratz in the Great Square of the Pegasus constellation. Or you could just put those phones to work and use one of your multitudinous *apps* to lead the way.

Human astronomers use constellations to divide your celestial sphere into regions with clear boundaries so that they can more easily describe an object's location in the sky. Their constellations differ from yours since you likely think of constellations as the pretty shapes you get when you play connect the dots with my brilliant stars. Your astronomers would call those asterisms, and some of them have unreasonably strong feelings about the distinction between the two terms.

That pesky IAU may have had the last word about what is and isn't a constellation, but they certainly didn't have the first. Claudius Ptolemy, a Greek astronomer who lived in the second century after your Christ, wrote about forty-eight constellations in a book he called *The Almagest*. They included the twelve zodiac constellations that trace out your ecliptic (or the path of the sun across the sky over a year), twenty-one constellations to the north (including Andromeda), and fifteen constellations to the south. Ptolemy didn't make up these asterisms himself, though. Like so many of the fraternity brothers that are his people's progeny, he relied on plagiarizing cleverer astronomers. The shapes were adopted from Egyptian, Babylonian, and Assyrian astronomers, though the exact outlines and assignment of stars to different constellations varied across both time and space. The Greeks affixed their own stories to the shapes, as did the other groups. But much again like the frat bros of your age, the Greeks were more successful in disseminating… their myths! Get your mind out of the gutter!

Ancient Chinese astronomers developed their own set of constellations over hundreds of years, separate from Greek or European influence. The precise number of asterisms shifted over time and according to which astronomers you consulted, but most would agree that there were a few hundred in the Chinese sky. A couple dozen more were added after the sixteenth century, when Chinese astronomers consulted European star charts and were able to see what the deep southern sky looked like for the first time.

They may not have agreed on the number of asterisms, but ancient Chinese astronomers all agreed on how many sections of the sky there were. They divided the sky into three enclosures around the north celestial pole: the Purple Forbidden, which is visible all year; the Supreme Palace, which is visible in the northern spring; and the Heavenly Market, which is visible in the fall. They also sliced the sky along the ecliptic into twenty-eight different lunar mansions (so called because your moon appears to live in each of these different slices for one day as it orbits your planet), seven mansions for each of four symbols: the Azure Dragon of the East, the Black Tortoise of the North, the White Tiger of the West, and the Vermillion Bird of the South.

The stars in the IAU's Andromeda constellation aren't assigned to any single Chinese asterism, but instead straddle the line between the Black Tortoise and the White Tiger.

Down in your southern hemisphere, the Incas had the best view of my prettiest features thanks to the sixty-degree tilt between my midplane and your solar system's that keeps your planet's southern half angled towards one of my arms.

They could see both my brilliant stars and my impressive gas clouds (which sadly just looked like darkness to their tiny eyes that lacked the ability to see infrared light), so they came up with two types of constellations: light and dark. Their bright asterisms depicted inanimate beings, mostly animals that nevertheless watched over their earthly counterparts. The dark constellations, however, like Yacana the Llama and Mach'acuay the Serpent, were said to be *alive* and drinking from the river that my spiral arms project across your sky.

Unlike Yacana and Mach'acuay, the Andromeda constellation and the sparkling galaxy it frames can be seen by everyone for some time every year. You should thank your lucky stars. Not every planet in my body is fortunate enough to have an unimpeded view of Andromeda, or, as life-forms in several billion years might know it, your stepgalaxy.

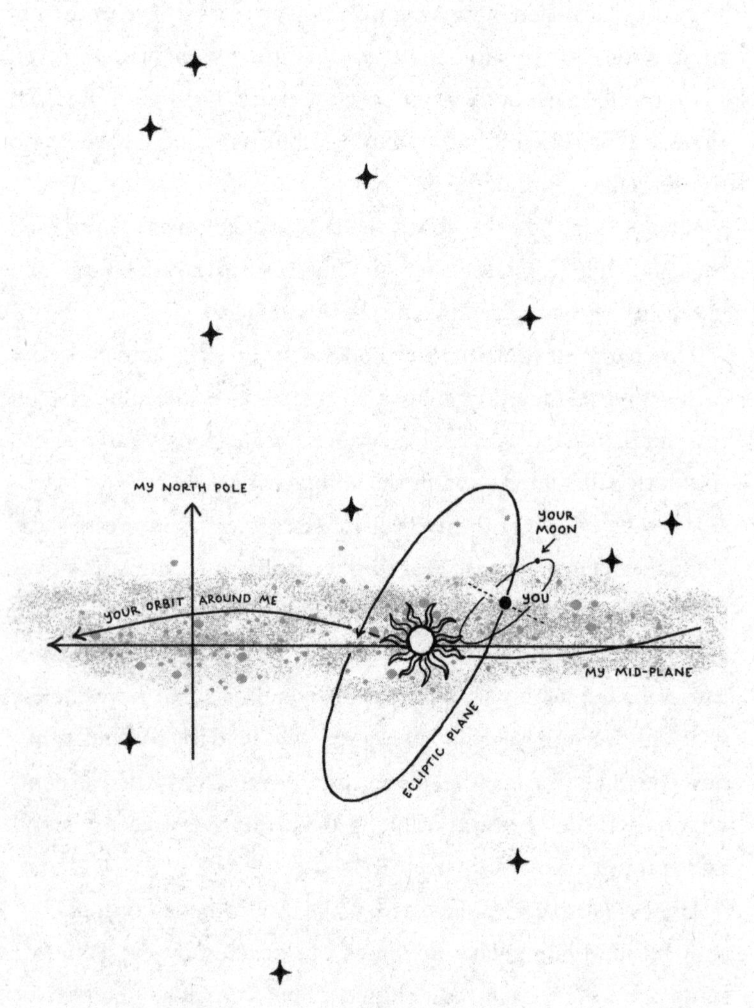
MY NORTH POLE
YOUR MOON
YOUR ORBIT AROUND ME
YOU
MY MID-PLANE
ECLIPTIC PLANE

CHAPTER TWELVE

CRUSH

THE FIRST TIME I NOTICED Andromeda was early in the universe, only a few hundred million years after the Big Bang, when there were substantially more of us galaxies packed into a space about a thousand times smaller than the universe is now. There was still plenty of room for all of us to go about our business—much of the universe's growth happened before it was cool enough for atoms, let alone stars and galaxies, to form; and most of us were considerably slighter back then before many thousands of mergers brought those of us who remain to our current sizes—but we still clustered together in small groups.

Your astronomers chalk that coalescence up to gravity, and they're not necessarily *wrong*. Galaxies use gravity in much the same way that you use your muscles: to move things around. But the ultimate reason that we clustered together back then was simply that we wanted to be near each other. We wanted to talk, to trade, to fight, and... do other things that would probably make you blush. Having blood suddenly rush to your face must be uncomfortable, so I'll spare you the details, but suffice it to say we were like you humans when you first go away to college, not yet ready to strike out fully on our own but prepared for some depraved experimentation.

Humans do this clustering on many scales. Even past those uncertain formative years, you continue to cluster together at parties and other social events when there's plenty of room to spread around, and you cluster en masse in your cities around Earth. I blame your urban centers for stealing your attention away from me, but as a galaxy who takes pride in my own shiny things, I can't deny that there's a certain beauty in the way they look from afar at night.

So, there we were, gathered in what, unbeknownst to us at the time, would eventually become the Local Group. And by some chance, some lucky coincidence that's almost enough to make me believe in fate, Andromeda was there, too.

When your astronomers look at Andromeda now, they see a standard barred spiral galaxy in the Local Group, currently a little less than 800 kpc away. With a diameter of about 70 kpc, or 220,000 light-years, Andromeda's luminous body is about twice as big as mine, and just a bit brighter (at all wavelengths, but you'll probably just care about the narrow range of the electromagnetic spectrum that your basic human eyes can perceive).

The system that human astronomers use to quantify how bright objects are was invented in Greece a little over two thousand years ago by a mathematician named Hipparchus. After ranking the stars he could see in your night sky by brightness, he split them up into six magnitude levels. You've probably heard of some of the brightest—humans call them Sirius, Vega, Rigel, Betelgeuse. But Hipparchus must have had either a wicked sense of humor or a very confused understanding of the concept of "more," because the system is backward! The brightest stars were called "first magnitude" and the faintest, "sixth." Your modern astronomers have decided to continue going along with Hipparchus's malarkey for the sake of *tradition* (a common human folly, it seems), but they've had to add more magnitude levels as the invention of telescopes opened their eyes to the billions of stars too faint for you to see, and as they discovered stars even more brilliant than Sirius, who seemed faint only because they were so far flung.

Most modern astronomers use the star Vega as a reference point and assign it a magnitude of 0. That's a perfectly respectable choice of anchor. Vega is bright enough for anyone to find, and in about thirteen thousand years, your planet's axis of rotation will wobble until it points towards Vega instead of Polaris. To scale the magnitude up or down by 1, divide or multiply Vega's brightness by a factor of 2.5. So the difference in brightness between a first and sixth magnitude star is 2.5^5, or about 100. This is done to preserve the relative brightness of Hipparchus's scale, because the faintest star your eyes can see is about 100 times dimmer than the brightest.

According to this ridiculous system, Andromeda has an apparent visual magnitude of around 3.4. That means that Andromeda is big and bright enough for even you to glimpse, if you know how to find it.

Your astronomers have vacillated back and forth on which one of us they think is more massive. Their measurements of Andromeda's mass have ranged from 700 billion to 2.5 trillion solar masses, but the most recent estimates favor the lower range. In your astronomers' eyes, that put Andromeda and me on more equal gravitational footing for a while, until more recent human measurements of my mass revealed a higher value than they expected. After analyzing the data from the Gaia mission that measured the motion of my stars, they think I'm closer to 1.5 trillion times the mass of your sun.

They can keep refining their mass calculations, but for once in my life, I don't care who's more massive. That matters only when your goal is overpowering every galaxy you meet, and I have no desire to dominate or control Andromeda. Besides, mass is just part of the equation. I may be more massive, but Andromeda has twice as many stars, around a trillion compared to my few hundred billion.

This zest for life and forming stars was one of the first things I loved about Andromeda, but don't let this apparent productivity fool you. There's still a greedy, destructive killjoy nestled in the center of the glamorous galaxy next door. Andromeda's supermassive black hole is much bigger than mine, around fifty million times more massive than your sun. I consider myself fortunate that Andromeda trusted me enough to share the source of these struggles over time. Those are not my stories to share, but I will say this: sometimes galaxies who appear strongest and happiest are the ones who are hurting the most.

From our first shared glance ten billion years ago, I could

tell that Andromeda was special, despite being much smaller and dimmer. I wouldn't say I have a *type* when it comes to galaxies—spirals and ellipticals are all beautiful in their own ways, and bars add some nice chaotic structure—but I have always found it hard to resist a galaxy with some *mass* to it. Nor was it only the impressive dark matter halo that Andromeda had already managed to collect, though I'd be lying if I said I wasn't intrigued by such a sure sign that this galaxy would one day be a mighty force to be reckoned with. It was more than just Andromeda's physical appearance that caught my attention.

I was pulled in by the way Andromeda moved through the group with a vigor and confidence that seemed to say, "Yes, I'm here and I want you all to know it, but I don't care what you do with that information because I don't need any of you!" A self-possessiveness that seemed to have its own gravitational force. I adored how Andromeda fed so unselfconsciously and without reservation on other galaxies. To this day, when Andromeda overpowers a smaller galaxy, it doesn't seem violent. If anything, it leaves the impression that other galaxies just *offer* themselves up to Andromeda. "Please, take everything that I am, for I am unworthy to exist in your presence."

Pathetic.

But relatable.

Of course, I wasn't the only galaxy in the group to notice the magnificent creature who moved among us. I was just the most patient. You see, other galaxies decided to—how do you humans say it these days?—shoot their shot right away. The cockiest ones approached when they were still little more than flimsy wisps of star-speckled gas, as if Andromeda would be desperate enough to accept the first galaxy who showed any interest. I'm not saying Andromeda is shallow, by any means, but a galaxy's got to have standards! And Andromeda's are too high to settle for any hot-aired blowhard who hasn't found or demonstrated its potential. What if it turned out to be prolate? Andromeda's too young for a galaxy with totally triaxial potential.[1]

I knew I had to do something to pique Andromeda's interest, so I picked a fight with the most colossal opponent I could find. Weighing in at 50 billion solar masses, Gaia Enceladus was one of the largest dwarf galaxies in the neighborhood, and wanted everyone to know it. GE sauntered around the Local Group, threatening to fight and eat everyone in its path, insulting their velocity dispersions, and greatly exaggerating its gravitational influence. Seriously, GE made Trin seem like Mx. Congeniality. It had to be stopped, and I was the best galaxy for the job.

Inevitably, I won the battle, and couldn't be bothered to keep track of Gaia Enceladus in the aftermath, but human astronomers, who didn't have the benefit of *witnessing* the titanic clash, pieced together the story from GE remnants scattered around my body. Most of GE's stars, gas, and dark matter dispersed to the point of being largely untraceable, but there are some globular clusters orbiting around in my halo, steadfast in their commitment to GE and each other.

Accomplished galaxies like myself carry within us globular clusters left over from a plethora of mergers—along with some globular clusters made from our own gas—so your astronomers first had to determine which originated with Gaia Enceladus.

To do so, they studied the ages and metallicities of the clusters' stars, and the dynamics of the individual clusters as a whole. It's ingeniously indirect of them. Gaia Enceladus's globular clusters—well, they're mine now, and have been for several billion years—are old, poor in metals, and carrying too much kinetic energy to have originated within me.

But lest we forget the purpose of this confrontation, let me say that my hopeful (but not desperate!) ploy for attention worked. I heard from Sammy, who claimed to have heard from Fornax by way of Phoenix, then Pisces, all the way back to Pegasus, that Andromeda was impressed by how I had handled my encounter with Gaia Enceladus and was interested in getting to know the galaxy who had rid the neighborhood of such an irredeemable bully.

Knowing Andromeda was open to my advances, it was time to make my move. But after watching galaxies throw themselves at Andromeda for too long, I couldn't rush into anything. The fruits of romance are far sweeter if given plenty of time to ripen, so I sent Andromeda a message the old-fashioned way: with stars, of course.

Humans have a history going back thousands of years—which for you is a long time—of tracking the motions of stars, but the only tools you had for most of that time were your gelatinous ocular orbs, and the farthest star your weak human eyes can see is a measly kiloparsec away. It's known as Cassiopeia, not to be confused with the constellation of the same name. My messenger stars hail from much farther afield than that old queen's namesake, so it's only in the last few decades or so that your astronomers have been able to systematically record the positions and 3D velocities of enough stars at a great enough distance to notice my stellar communiqués. Though, they couldn't have known that I was using the stars to send love notes to another galaxy.

These messenger stars must move hastily enough to escape my gravitational pull so that they can reach their extragalactic recipients. So briskly, in fact, that human astronomers call them "hypervelocity stars." Hypervelocity stars needn't travel all the way to another galaxy to pass along the message, though; as long as the star has left the sender galaxy, the recipient should be able to decipher what it says.

Your astronomers discovered their first hypervelocity star in 2005 and named it SDSS J090745.0+024507, an adorable pet name. From my resting perspective,[2] it's moving a bit faster than 700 km/s, which puts it well above my escape velocity of roughly 550 km/s. Since that first discovery, your astronomers have found a few dozen more, along with about a thousand "high-velocity" stars that haven't reached escape velocity but are still moving significantly faster than the other stars in my disk. Those are drafts that I never sent, because I couldn't share anything less than my best with Andromeda.

At first, human astronomers assumed that stars could reach these high speeds—some of them an appreciable fraction of the speed of light—only if they got a boost from the immense gravity of a black hole. If they had just bothered to get to know me, the real me, just a *little*, then they would have known I wouldn't use that monster to send love notes. But they discerned as much on

their own when they discovered a messenger star in 2014 called LAMOST-HVS1 zipping away from a point firmly in my disk, far away from Sarge. I typically use the gravity of a massive binary pair to send my notes, but any sufficiently energetic dynamic system will do.

Not all my hypervelocity stars are going to Andromeda. Some of them, like SDSS J090745.0+024507, are formal correspondence to other galaxies with messages far too dull to repeat here, but pertinent enough that I wanted them in writing. Others are congratulatory messages that I was obligated to send whenever another galaxy felt it had achieved something, like when Leo made its millionth star. A few of them were pranks I sent to Larry. You've no doubt gleaned my impeccable sense of humor, so it should come as no surprise that I have a bit of a reputation as a jokester around the neighborhood.

I knew the first star note I sent to Andromeda would set the tone for our entire relationship, which even at the time I hoped would be a long one. It had to be clever and sweet—but not saccharine—and straight to the point because Andromeda doesn't have patience for galaxies who waste its time. I spent a few millennia crafting the note, packaging it inside the most pristine F star I could make, and I waited...

And waited...

And waited a hundred million years for a response. I know that's not very protracted in the grand scheme of cosmic time, but even humans recognize that waiting, just like black holes, can bend and stretch time for anyone unlucky enough to fall into its clutches. I tried to throw myself into my work as a distraction, but it would have been obvious to anyone paying attention that my mind wasn't on the task at hand. I made so many brown dwarfs...

But Andromeda did eventually write back, thank the Cosmos! I won't reveal the details of our private correspondence—a gentlegalaxy never does—but we've been messaging and drifting towards each other ever since.

Recently, in the last few million years or so, our halos have started to touch. Not our stellar halos. No, nothing quite that intimate yet, but our circumgalactic halos overlap in a few exciting spots. When my massive stars die in powerful supernova explosions, they push gas and dust away from them. Some of that material, along with a fraction of the matter that Sarge flings away from my center in an effort to keep me from easily forming stars, collects in a giant cloud around me. The cloud is clumpy, denser in some parts and more diffuse in others. Much of the cloud is heated up by radiation from my stars because they toil so arduously.

Andromeda has one of these halos, too, and both of ours extend out more than a million light-years in almost every direction, far enough to overlap. It's like that brief, tantalizing moment before you touch someone for the first time, when you're just sharing the same air and the future is full of possibilities. I could bask in the glow—not a literal glow, mind you, because gas in the intergalactic medium is pretty cold—of this feeling forever. Or at least for a couple billion years, which for you is basically the same thing.

But I can't say my relationship with Andromeda has been without its hiccups. At times, I wanted to joke that I felt like I was courting a galaxy called An-drama-da, but I didn't think it would be very well received. Much of the tension came about because of Andromeda's trysts, which I tried—and failed—to ignore. It's not that I resented Andromeda for interacting with other galaxies, even merging with some of them—who am I to judge a galaxy for doing what it needs to get by?—but it's never pleasant to see the galaxy you love entangled with someone else.

The first of these significant mergers happened many billions of years ago when Andromeda was still young and actively growing. Naturally, it was a short-lived relationship. Even back then, Andromeda struggled to find an equal, someone who could hold their own in the turbulent process that is any galactic collision. Andromeda grew from the experience (literally) but remained unsatisfied. Now, the only evidence that can be found of this brief encounter is a small collection of globular clusters orbiting in Andromeda's outer halo. Human astronomers—the ones who call themselves galactic archaeologists—have studied the motion of these globular clusters and can tell that they're the result of such a long-ago merger because they're spread in a way that can only be achieved after drifting apart for several orbital periods. Orbital periods around entire galaxies like Andromeda and me can take hundreds of millions of years.

Maybe four billion years ago, not that I'm counting, Trin made its move on Andromeda. I can't really be sure who struck first, but I have a hard time imagining that Andromeda would have initiated. After all, we were exchanging some extremely flirty messages at the time. But Trin and Andromeda did have a close encounter—it never advanced to an actual merger—that triggered rapid star formation in each.

You might expect me to feel jealous over this little rendezvous. How utterly human of you to assume that I wouldn't celebrate Andromeda's happiness. Though I will admit to a small petty spark of joy when Trin's star formation rate was about ten times higher than Andromeda's after the encounter. Clearly, one galaxy enjoyed the interaction a lot more than the other. I'm not at all surprised that Trin's performance left something to be desired.

Most recently, about two billion years ago, Andromeda had a serious merger, or at least more notable than Andromeda's previous galactic encounters. The galaxy must have done something right, because after the collision, Andromeda churned out stars at its most rapid rate ever. Nearly one-fifth of all of Andromeda's stars were produced in the afterglow of that coupling. But even from the other side of the Local Group, I could tell it was just a passing fling. The other galaxy was closer to Andromeda in size and power than the others had been—about a quarter of Andromeda's mass—but not close enough. And just as I expected, a few million years later, Andromeda was still on a direct path towards me, and the other galaxy was left as a small core of its former self, relegated to the position of satellite galaxy and destined to spend the rest of its eternity orbiting the ex who chewed it up and spit it out.

Some of you might see this as a sign that Andromeda is cold and careless; I've always seen it as a sign of Andromeda's aplomb and conviction. Plus galaxies like to have fun, too.

Human astronomers never deigned to give that interloping galaxy a name, but they call the sad little core left behind M32.

Throughout all of this, I've been spared any pangs of jealousy thanks to (1) my superior intellect and emotional maturity and (2) my understanding of the distinction between what human astronomers call "minor" and "major" mergers.

Minor mergers occur between two galaxies that are unequal in mass or power or responsibility. They are more common and less consequential than major mergers, though they are responsible for spurring much of the growth and star formation that we galaxies accomplish in our long lifetimes. My minor mergers—and the introspective work I've done in between them—have made me the galaxy I am today. I'm sure Andromeda feels the same way. But it's important to recognize minor mergers for what they are: temporary. Only a fool would expect something so uneven in nature to last forever.

A major merger, on the other hand, is a true partnership between galaxies, one that permanently alters both—or all, as some galaxies feel more content with multiple partners—parties in ways that are difficult to precisely predict. Most major mergers happen between galaxies who share nothing more than an instinctual, gravitational bond. Indeed, your astronomers only describe major mergers as a collision between two galaxies with a mass ratio close to 1. So cold and quantitative. They totally ignore the quality of the merger, which depends on much more than gravity.

Don't get me wrong, Andromeda and I have plenty of that, too. Our gravitational bond draws us together at a speed of about 100 km/s. I'm sure that sounds fast to you, but it feels *achingly* slow to me. Like I'm watching myself in one of those human movies where the romantic interests run in slow motion towards each other, and I just want to scream, "Hurry up! You're wasting time!"

As Andromeda and I get nearer, our respective gravitational pulls will get stronger, and we'll jerk[3] towards each other, spurred by the knowledge that we're just that much closer to *finally* meeting.

But it's more than just our similar masses that's drawing us together. Andromeda and I have been building and strengthening our bond for ten billion years. I've lost count of the messages we've sent back and forth to each other, though I treasure each and every one of them. Slowly, over time, they became more intimate, revealing more of our inner selves—our fears and hopes and petty grievances. We divulged the shames and insecurities that fueled our black holes, and helped each other heal those old wounds. The black holes are still there, but it's easier to ignore them now.

What started as fascination carried over the galactic rumor mill, so to speak, has since blossomed into deep affection and mutual respect. We still have to wait about four or five billion years before we can meet, though. If there are any humans left by then, they'll be able to see Andromeda's effulgence in more detail with their own eyes as the spiral beauty will stretch across half your sky once it gets close enough.

When we do converge, we're not just going to crash and stick together. We'll test each other out first, see if we have the same chemistry in person that we do in our correspondence. You wouldn't marry someone from the internet the first time you met them in real life, would you? Actually, don't answer that. Besides, sticking together in one pass isn't even how galaxies move. We move on orbits, dancing around each other. So, after that initial collision, Andromeda and I will pass through one another—

probably more than once, getting closer each time—and then swing our way back into each other's cool embrace.

Your astronomers have tried to model that dance in advanced computer programs. (Amazing what secrets of the universe your computers can unlock when you're not using them to watch filthy videos you don't want your friends to know about.) According to their simulations, the merger will take about six billion years to complete—and some of you have the audacity to complain about the length of *your* marriage ceremonies! The commotion and the mixing of our gases will trigger a period of rapid and intense star formation. Some stars will be ejected, but that's just us sending announcements of our union to our dearest galaxies (and to those whose smug faces we want to rub our good news in). After all is said and done, we'll be a new galaxy! An elliptical, actually.

We'll have to learn how to move together as one in our new body. All galaxies move ceaselessly to avoid being crushed under our own weight, but ellipticals move differently than spirals. Your astronomers would say that a galaxy's kinetic energy from motion must equal its gravitational potential energy. Whereas spirals like Andromeda and I prefer to keep our motion ordered with stars and gas that mostly rotate on circular orbits, ellipticals are supported by random movement. It's just too hard to stay as rigidly organized as you used to be once you merge your life and affairs with someone else's.

Our black holes will have to learn how to live with each other, too, and neither one of them will like the new arrangement at first. But after several billion years of orbiting each other and siphoning off energy—both negative and kinetic—to the stars and gas around them, they'll realize that they can torture Andromeda and me better as a team. They'll spiral towards each other, and the collision will send out gravitational waves that warp space-time for more than two million light-years. It's just like a black hole to unload problems on other people and then make a big spectacle of itself. Your astronomers have never seen two black holes in their final stage of colliding, but they have detected the ripples that result from their mergers. They're not certain how the monsters dissipate that last amount of energy in the near-frictionless final parsec of their encounter, a mystery they call, rather appropriately, the final parsec problem.

Make no mistake, Sarge and its new friend will merge; it's just a matter of time. I'm not worried they'll cause much trouble, though, not if Andromeda and I get to spend the rest of eternity distracting each other with stories and dancing. Andromeda always wanted a partner it could dance with.

Apparently, it's common for your silly human media to make up couple names for celebrities in relationships, and I'm willing to indulge. If you go by number of mentions throughout history, I'm certainly the biggest celebrity humanity has ever known, so of course Andromeda and I have a couple name: Milkomeda. It has a nice ring to it, doesn't it?

Your astronomers are quite invested in my relationship with Andromeda. I've watched—not actively, I just have a wide attention span—many human astronomers go through their training. Since you learned about our imminent collision sometime in the 2000s, one common exercise for fledgling human astronomers is to estimate how long it will take for Andromeda and me to meet. I like to believe that the early exposure to our epic romance fosters a much-needed sympathy for galaxies in their impressionable young hearts and minds.

Trainee astronomers are also asked to estimate how many stars will collide when Andromeda and I do. What a human question to ask! Only someone with an extremely limited view of the universe, someone totally unfamiliar with life on macroscopic scales, would wonder such a thing. It's hard for humans to really consider all the necessary factors (e.g., our respective stellar spatial distributions, our angle of approach, gravitational influences from other galaxies like Trin who will probably rudely try to hold Andromeda back), but with a few simplifying assumptions, you'll see that probably only a few stars of ours will ultimately hit each other. I don't know why you're all so worried about it. By the time Andromeda and I merge, your planet will likely be long gone.

Huh, isn't it kind of funny that just as humanity as we know it is ending, my life with Andromeda will just be beginning? No offense, but I'm kind of excited! It's nice to have something to look forward to.

CHAPTER THIRTEEN

DEATH

I'M WRITING THIS BOOK TO share my story with you, a new generation of humans who could display greater respect for the galaxy who works so hard to keep a blanket of stars over your heads. I don't know if you excel at picking up on context, so I should warn you that we've passed the part of the story where I tell you what *was* and are now squarely in the part where I look to what *might be*. So here is the last thing I can tell you with certainty: I am going to die one day.

I know that might shock you. It's difficult to imagine a universe without me, and it must be near impossible for you to imagine a force strong enough to wipe out a galaxy as powerful, resilient, charming, and humble as myself. If Sarge couldn't knock me down, what could? But I can guarantee that my days are numbered—though thankfully my number is much, much, much larger than yours.

How often do you think about your own impending death,

human? About the fact that one day, no matter how hard you try to fight it, the little organic machines keeping your body running will seize up, and your fleshy form will rot and decay to the point where only the most astute of your scientists would even be able to recognize what's left of you? And that's only if you don't get burned to ashes or shot off into space or chewed up and swallowed by one of those fascinating predators on your planet. Surely you must take some comfort in the fact that the people you leave behind will most likely go to great lengths to honor your memory. Oh, the parties that have been thrown to celebrate your sweet, short human lives! Have you envisioned your future corpse's party? I'm sorry, your *funeral*? Sammy tells me death is a sensitive topic for you humans. Have I written about your inevitable demise with appropriate tact?

I used to think about death all the time. Not the unseemly decomposition part, but about destruction more generally, about the moments that separate something's presence from its ultimate absence, moments that could last seconds or millennia. I think about it less now, and when I do, the death I imagine is far, far in the future and I'm not afraid of it.

Why would I be?

It makes sense that you humans would be afraid to die. Giving up your favorite activities, leaving your loved ones behind, and venturing into the ultimate uncertainty must be frightening to a creature so shortsighted and ignorant. (No need to take offense, your species just objectively isn't very advanced.) And it seems like anything can kill you. Falling, stabbing, burning, somehow both drinking too little water *and* too much?

Galaxies don't have any of those fears. I'm not burdened by the same uncertainty as you. Andromeda and I are going to merge so completely in just a few billion years that we'll essentially become one galaxy. We'll die at the same time, and I won't have to worry about leaving anyone I care about behind. I know, thanks to the work done by researcher galaxies throughout the universe, exactly how it's all going to end. Okay, maybe not *exactly*. We galaxies are intelligent, not magical, and even our science comes with some uncertainty.

Don't worry, I won't spoil the ending for you, though I will tell you what your human scientists think could happen. Since your experience with death is limited to fragile organic beings, I should explain what it really means for a galaxy—a massive, self-sustaining system with a disparate but coherent consciousness—to die. As the threshold of birth is a little blurry for us, you'll understand if our death is similarly ambiguous. One thing is for sure: my death will be unmistakably more consequential than yours, and it will be more spectacular than any of your imagined doomsdays.

Sometimes, death for a galaxy can mean losing control. Actually, most of the time, since the overwhelming majority of all galactic interactions throughout cosmic history have been minor mergers where one galaxy completely overpowers the other. You'd think the issue with those ruinous encounters would be the fact that the smaller galaxies are ripped to shreds and have their—for lack of a better analogy—guts strewn carelessly about. But the truth is that the little galaxies that I've destroyed could technically be put back together if someone cared enough to painstakingly comb through every square parsec of my body to pick out which stars and blobs of gas belonged to them. (Though, by now, I'm sure I've already gobbled up most of the gas to make new stars of my own.) No, we galaxies are prideful beings, so the real tragedy in these scenarios comes from the loss of agency.

Of course, there are also galaxies who die more traditional deaths. Some human astronomers call old ellipticals "red and dead" when they stop forming stars. I think it's more reasonable to consider a galaxy dead when its stars have actually died, not just stopped forming. We are, after all, just collections of stars and gas and dark matter packed close enough together to be gravitationally bound. That kind of death would take a loooong time, even from a galaxy's perspective, but would be my preferred way to go. Just... fade slowly into the eternal night.

These kinds of deaths are trivial—maybe even downright temporary, since any galaxy can be reborn with a little infusion of gas—when compared to the inevitable end of everything that has ever existed (and will ever exist). That's right, I'm talking about the truest doomsday there is: the end of the universe!

Over the last human century, your astronomers have come up with several widely regarded hypotheses for how the universe might end, and only one of them is so wrong that it becomes comical. Charmingly named to evoke the Big Bang, they are, in no particular order: the Big Freeze, Big Rip, Big Slurp, Big Crunch, and Big Bounce (this last one is a sneaky bonus riding on the coattails of the Big Crunch, but I applaud the thoroughness in their distinction).

Before I tell you about the best guesses of your species' leading minds, you should know that—for the most part—the ultimate fate of the entire universe depends on two factors: the average density of our observable universe and the long-term and large-scale behavior of the expansive force your scientists call dark energy.

The density—of both matter and energy, since the two are interchangeable—can be broken down into three components: density of matter both baryonic and dark, density of radiation from relativistic particles (photons and neutrinos), and density of dark energy.

As the universe has grown and evolved, different components have dominated the density. At first, during the brief period of dramatic inflation, the universe was dominated by an expansive energy that your scientists attribute to the inflaton quantum field (more on those in a bit). Immediately after inflation, when the universe was so hot that atoms couldn't even form, the density was dominated by radiation for about fifty thousand years. Next it was matter's turn to reign supreme for about nine billion years. Now that the universe has expanded enough that the matter is sufficiently spread out, we're most heavily influenced by dark energy.

The density of the universe at any given time can be compared to the critical density from chapter 6, the Ω symbol. A Soviet mathematician defined the critical density in the 1920s by assuming that the universe, that fabric of space-time that you humans love to imagine so much, is totally flat and can expand outward in every direction for an infinite amount of time.

That mathematician's name was Alexander Friedmann, and he was the first person to publicly oppose Einstein's belief in a static universe. Einstein, that smug bastard so famous among you humans that I don't even have to say his first name, was the first human to figure out black holes, you know. And instead of outing them as the cruel monsters they are, he made them sound cool! He thought he figured out *everything* with his little relativity theories, and yes, he got many things right. But you know what? He wasn't that brilliant, he was just lucky! He was wrong about the universe being still, and Friedmann was brave enough to disagree. It's a very small and specific bravery, but it counts, nonetheless.

Einstein had published his theory of general relativity, the equations that describe the behavior and consequences of gravity in our universe, including equations about how gravity affects its growth, or lack thereof. Friedmann figured out his own solution to—or maybe his own interpretation of—Einstein's field equations. It was an equation—everything's always an equation with these people—that described the size of the universe over time in terms of its density, curvature (its shape), gravity, and expansion rate. The expansion rate of the universe is typically denoted by your scientists with a capital H, and they call it the Hubble parameter. You may have heard the term "Hubble constant," and that is just the expansion rate of the universe *right now* because your scientists' egos plus your species' short life span means they inflate the significance of the present. If you know the Hubble parameter—and your scientists are confident (maybe too confident) that they do—you can solve for density. And the critical density is just what you get if you set curvature to 0. See, it's really not that complicated when you do away with all the math you humans invented to explain it. Friedmann's contrary calculations revealed that this theoretical critical density is 10^{-26} kilograms per cubic meter, which is roughly 10 teeny tiny hydrogen atoms in each space as big as one of your standard hot tubs. (Relaxing in a hot, bubbly soup happens to be one of the few human sensations I wouldn't mind experiencing.)

Sometimes, your scientists talk about the shape of the universe, which is absurd because the universe's shape is just a proxy for its density, and most of you humans are totally incapable of imagining a "saddle-shaped" universe. These aren't mere three-dimensional shapes, so you wouldn't be able to see them with your eyes, anyway. Don't bother trying to imagine what they look like unless you want to give yourself a headache.

The density of the universe is either less than, equal to, or greater than the critical density. Duh. That means that Ω, the ratio of the actual density to the critical density, can be either <1, 1, or >1.

If $\Omega = 1$, then the universe is flat by definition, because the critical density is the density of a flat universe. Once inflated, a flat universe will continue to expand, but it will be slowed to a halt by the sticky force of gravity after *precisely* infinity years. Keep in mind that Friedmann, as brave as he was, wasn't canny enough to include dark energy in his models. Only matter and relativistic particles contribute to the density in a simple Friedmann universe. With dark energy, it is possible to have a flat universe that never stops expanding, even after infinity years.

The universe would also expand forever without pause if $\Omega<1$, because in a low-density universe, there isn't enough matter/energy for gravity to halt the universe's expansion. Your scientists call this an "open" universe, and they say it's shaped like a saddle. That leather seat fitted to a horse's back that you make them wear so you can ride them? You humans are so bizarre.

If $\Omega>1$, the universe has so much matter/energy that dark energy can't overcome gravity's relentless grasp. This is known by human scientists as a "closed" shape, and in a closed universe, expansion slows to a halt and then reverses on itself, like a rubber band stretched to its limits.

Each of these possible density scenarios suggests different ways that the universe as we know it could end.

Your human scientists are confident that they've calculated the critical density correctly, and that they know the current average density based on wide-field observations. I think they should be a little less sure in their density values, but at least we can all agree that your scientists don't understand anything about dark energy, especially on the sizable scales that your kind has always struggled with. And that means your scientists—or at least the good ones—are the first to admit that they don't know how the universe will end. But I suppose it's worth telling you their ideas, if only so you can sound smart by talking about me—and them—at dinner parties.

BIG RIP

In a sparse open universe with strong enough dark energy (or a flat universe with even stronger dark energy), the universe will continue to accelerate in its expansion until gravity is too weak to hold things together, even on the scale of individual galaxies. Your scientists call this scenario the Big Rip, and it's a fitting name for what is my least favorite human-proposed end of the universe.

Right now, dark energy is still too weak to overcome gravity on small, local scales. It can expand the space between galaxy clusters, and it has. It's why I haven't seen some of my childhood friends in billions of years, and probably never will again. But gravity is still strong enough to hold clusters, galaxies, stellar systems, and stars together.

But as the universe expands and the matter expands with it, my trusty tool gravity will become less effective at holding it all together. First, the galaxies will start to move away from each other within their clusters. It's always a shame to see communities torn apart, but the real problem will come when individual galaxies start feeling stretched. This is far worse than when I tear apart little dwarfs, because at least then parts of the galaxy stay close together, maybe passing each other every once in a while as they orbit the galaxy who devoured them. In the Big Rip, the space between individual stars and gas particles will expand. Andromeda and I will be irrevocably separated, a crushing blow after such a long courtship. To add insult to injury, the space between atoms in our stars will stretch apart. Even the strong nuclear force, 6×10^{39} times stronger than gravity, will buckle in the face of dark energy, and individual atoms will be torn apart.

Whether or not all this happens depends on the density of the universe and the nature of dark energy, but more specifically, it depends on the ratio of dark energy's pressure to its density. In other words, how much push do you get from the amount of dark energy in a given volume of space? If dark energy's push is relatively weak, then dark energy will dissipate over time and the Big Rip won't happen. Andromeda and I would get to live out our extended happily ever after! But if dark energy's push is strong, then our days together would be numbered.

How numbered, though? Well, that depends on the strength of dark energy's push, of course, as well as the universe's expansion rate and the density of gravity-inducing matter. In your astronomers' most realistic bad-case scenario, the Big Rip would happen in just about twenty billion years, which simply isn't enough time to spend with the galaxy you love.

I really don't like to dwell on this scenario, so I have no qualms telling you that they don't think the Big Rip is very likely. Your scientists' admittedly crude measurements of dark energy's strength and the universe's current density point towards less devastating ends.

BIG FREEZE

If the universe is allowed to expand forever without any of that dramatic and totally unnecessary cosmic ripping, then it could end in the Big Freeze, aka the Big Chill. Those who like to laugh in the face of convention and uniformity also sometimes call it the universe's Heat Death. The Big Freeze could happen in an open or flat universe with or without dark energy, and in a closed universe with strong enough dark energy.

The universe gets colder as it expands, and that trend will

continue into the far future. If the universe expands enough and all the matter gets spread out and particles slow down, then the average temperature of the universe will reach 0 K, or at least get very close to it. In this scenario, Andromeda and I would have enough time to gobble up all the other galaxies in the local group—sorry in advance, Sammy!—and become one big megagalaxy. Together, we'd continue to form stars until we ran out of usable gas, which should be in about a trillion years, maybe a hundred trillion if we're lucky. Isn't it poetic that as I form the last of my stars, the low-mass survivors from my first batches will die? And that final generation of stars will fade slowly into a cold, dark oblivion until a lonely M dwarf wrings the last bit of light from its exhausted core. And with all the other galaxies and clusters pulled out of sight by dark energy, beyond a line your scientists call the cosmic light horizon, there would be no new gas to keep us going.

You might think your human scientists would be content to stop there, that they wouldn't wonder about the goings-on of a dark and dead universe. You would be wrong, because human curiosity is notoriously insatiable.

It was yet another Soviet mathematician in the 1960s who proposed the idea of proton decay as a way for matter to break down at one of its most fundamental levels. Protons are some of the most stable particles because they're already so light (the lightest in the category of particles that your scientists assigned them to) that there are only a few types of particles they could decay into, like a positron or a meson. But Andrei Sakharov wasn't intimidated by the proton's stability. To him, the proton was another particle, and particles decay. The idea was totally theoretical

when Sakharov came up with it, and your scientists still haven't been able to observe proton decay in action. In their defense, protons should take decillions of years to decay naturally.

If proton decay is real—and I'm not telling how big that *if* is—then even the protons that make up my long-dead stars will decay away into tiny useless particles. Only black holes will remain, which, when deprived of any galaxies to torment, will evaporate away as they slowly lose energy in the form of what your scientists call Hawking radiation.[1] Hawking radiation is a story for another time, but suffice it to say that it is yet another theoretical phenomenon that would drag out the ultimate end of the universe by...well, by an amount of time that even I have a hard time imagining: 10^{100} years for the most massive of them. That's ten duotrigintillion years! Of course, black holes would insist on having the ultimate last word.

BIG CRUNCH

A matter-dense universe ends just like it begins: with a bang.

In this scenario, the universe expands until the force of gravity from all the dark and baryonic matter slows the expansion to a halt, then reverses it. Your scientists put a lower limit of about sixty billion years between now and the potential Crunch, but it would likely take much longer. While the universe contracts, galaxy clusters will be pushed into collisions instead of getting farther apart like we're used to. Then individual galaxies will be forced to collide, though many of us would have already done that by then.

The scariest part of the Big Crunch for humans would likely be when space becomes so tightly packed that stars begin to collide, which has never really happened before. I've always tried to make sure that my stars have plenty of room to breathe—figuratively, of course. Luckily your kind will be gone long before that happens.

If the universe cools as it grows, what do you expect it to do as it shrinks? Obviously, it's going to heat up. The ambient temperature of the universe will increase from its current ~3 K probably up to the 10^{32} K it was around the time of the Big Bang. As the temperature rises, we'll reach a point where the universe is hotter than stars—M dwarfs at first, but the O stars won't be spared. The hot temperatures outside of stars will excite the gas molecules they're built out of, and the stars will literally boil away like a pot of water you forgot on the stove.

Just as there was a time in the early universe before atoms could form, the Big Crunch scenario leads to a time when the universe will be so hot that atoms break apart into free-floating protons, neutrons, and electrons.

This theory was more popular among your scientists in the latter half of the twentieth century, but it lost credibility in the 1990s when you realized the universe's growth was speeding up. Before the idea fell out of favor, it was staunchly supported by a man named John Wheeler. The very same John Wheeler who first used the term "black hole" at a scientific conference in 1967. He must have made a habit of backing bad takes.

Still, the Big Crunch could be kind of fun. It would provide a plum opportunity for Andromeda and me to grow closer, and we'd have a long life together before everything grew too hot for us to handle. And as the universe contracts, galaxies I lost touch with long ago would come back into view. Sure, there were a few galaxies that I was glad to see disappear beyond the cosmic light horizon, but who knows? Maybe they've grown over the eons just like I have. Or maybe the most annoying ones were destroyed. Either way, I can think of worse ways to spend the last million years of my life than reuniting with old friends.

If the universe's matter is smooshed into such an inconceivably small space, it could produce a gargantuan black hole the likes of which none of us have ever seen. Can you imagine the negativity? But the Big Crunch could also just be the precursor to something even more interesting, which brings me to the...

BIG BOUNCE

All too often, whenever one of you humans learns about the Big Bang for the first time, you ask the same predictable question: "But what came *before* the Big Bang?" The Big Bounce hypothesis offers up one potential answer.

The bounce in questions happens after the Big Crunch, when, instead of stopping as a black hole or an inert point mass, the universe rebounds and starts to expand again. In this scenario, the universe is in a continuous cycle of expanding and contracting.

The Big Bounce was popular among Einstein and his contemporaries, including a Belgian physicist—and Catholic priest—named Georges Lemaître. Lemaître was just a little slower than

Friedmann to solve Einstein's field equations with a description of a dynamic universe (i.e., one that is actively expanding or contracting). Though he is often credited as the first person to theorize that the Big Bang happened and that the universe sprang forth from a single "primordial atom."

You might be wondering, can the Big Crunch of a cyclical universe like this really be considered death? Well, maybe it's not an end to the universe, but there will never be another Milky Way or Andromeda. The universe wouldn't be able to re-create the exact course of events that led to our formation. As simple a system as you are, human, there will never even be another you, either.

Is this what it's like to imagine a future where other things exist but you don't? Well, I take no pleasure in it, so let's move on.

BIG SLURP

There is a universe-ending scenario proposed by some of your more cautious scientists that doesn't involve the density and expansion of the universe at all. At least, not directly. Instead, they say the universe could be "slurped" into the vacuum of a new universe because of a sudden and unpredictable shift in the fundamental laws of physics. Whether or not this happens depends on the stability of what your physicists call the Higgs field...which I'm sure I'll have to explain to you. Sigh.

According to most of your physicists, our universe is made of and influenced by small bundles of energy called fundamental or elementary particles. These specific particles can't be broken down into smaller parts, as far as your physicists can tell. The particles they've found so far are grouped into matter particles (electrons, all the quarks and neutrinos, etc.) and force particles (photons, gluons, the bosons, etc.) that carry the electromagnetic, weak, and strong forces. Your scientists discovered the first one in 1897 (the electron, thanks to a rather accomplished physicist named Ernest Rutherford), but now it seems they're finding new ones all the time, especially with that large ring they have in Switzerland where they smash hadrons together.[2]

Each of these particles is just a long-lived, discrete energy spike in its own so-called quantum field. I feel the need to stress that these are not actual, physical fields that you can directly manipulate in any way. They're a convenient and constructed mathematical arrangement, a hypothetical medium for transferring different types of energy around the universe. It might be easier for you to think of these fields as software programs running on the back end of the universe. There's a program, or field, that describes and controls electrons, a program for muons, another for inflatons, and so on. These programs rely on each other, so they interact in a way where, if you change or perturb one, it may or may not affect one or more of the other programs. Switching back to your scientists' language, they would call these interdependent software programs "coupled fields."

Human physicists are still actively trying to ascertain how most of these fields work and how they influence each other. One thing they're confident of is that they can fluctuate unpredictably, making it difficult for the energy spikes known as particles to last, like one of you fools building a sandcastle during an earthquake.

There are no fluctuations when the field is in an energetically stable state, and the most stable state of all is one with zero energy, also known as a vacuum state. That's why some of your scientists call the Big Slurp "false vacuum decay."

A field in a false vacuum state can seem stable, but that stability is a lie. At any moment, the field could drop to a lower state where the particles that represent it are long-lived but behave in a totally different way from our particles. That change, which could affect any number of other fields, would ripple through the entire universe at the speed of light so that we couldn't even see it coming. To return to my analogy, what happens to your computer when one of its core programs is rewritten? It reboots. So, if any of the fields governing our universe were to jump to a different energy state, the universe as we know it would cease to exist.

You do still use computers, right? It's so exhausting to keep up sometimes.

Your physicists think, with a healthy amount of uncertainty, that most of the universe's quantum fields are safely in their zero-energy state. All of them but—and now we're finally back to where I wanted to start this story—the Higgs field.

The Higgs field is associated with the Higgs boson, which some of your scientists call—either reverently or cheekily—the "God particle." The Higgs field is especially interesting (to both me and your scientists) because its interactions with other fields is essentially what determines their particles' masses. Protons, which are made of quarks, are heavier than electrons because the quark field interacts more strongly with the Higgs than the electron field does. Remember, a galaxy is useless without its mass.

Your scientists have been trying to measure the energy of the Higgs field, and they think it might be in a false vacuum state. That means that our universe could be headed for a reboot with no way to prepare for it.

Luckily, it would take either a lot of energy or an incredibly rare instance of what your scientists call "quantum tunneling" to push the Higgs field—or any other—into a new state. And when I say "rare," I mean the chances of it happening before my now and future stars are dead are basically negligible.

But it seems like a fast and painless enough way to go, so as long as it waits for Andromeda and me to finally connect, the Big Slurp can come whenever it wants.

Human astronomers are very confident that the density parameter Ω is very close to 1, but they don't know if it's slightly more or slightly less. Although the Big Freeze is most consistent with the observations of an accelerating expanding universe, you should keep your mind open to any possible ending. It was only one century ago that the supposedly great Albert Einstein thought the universe was standing still.

Your current scientific understanding of the world could be turned on its head at any moment. There could be another end of the universe coming, just waiting for its own Big title. Progress towards better understanding, even of something as unappealing as the end, means the science is working.

CHAPTER FOURTEEN

DOOMSDAY

YOUR KIND MAY HAVE STARTED theorizing about the end of the universe only in the last hundred years or so, but humans have been asking questions and telling stories about The End—often of human civilization or your entire world, which, since you didn't discover another planet until your 1780s, was pretty much your entire universe—for millennia.

Every time a new human civilization sprang up, they found that predicting the finale is a lot harder than trying to explain how everything started because you don't have the benefit of knowing the end result. Like your modern scientists, many of those early civilizations freely admitted that they didn't know how the world as they knew it would end. Some assumed that the ending would simply be a reverse of the beginning. But there were a few who claimed to identify, thanks to the foresight of gifted prophets, not only how the world would end, but also what would cause it.

Too many of these stories have been lost to time, either never written down or inscribed on a surface too fragile to survive your world's caustic environment and violent acts (*cough* book burning *cough*). But some endured to be told—and even believed—today.

One of the survivors comes to us from your old Norse folk in the tale of Ragnarok, meaning "fate of the gods." Many of you are probably familiar with the story, thanks to one of those ridiculously charming Australian brothers, but the original tale had more battle cries and fewer rock anthems.

As the Norse used to tell it, the end of the world starts with Fimbulwinter, a series of three harsh winters in a row without any summers in between. As food stores dwindle, humankind will forget that it was social cooperation that got you this far, and you'll turn on each other, killing not for glory in battle, but for greedy survival (which, confoundingly, is worse in the eyes of the Norse gods). A pair of wolves will eat the sun and the moon, and the stars will disappear to plunge your kind into darkness, as if something as powerful as my luminance could be snuffed out by a couple of ornery canines. But this is just the beginning of the end.

Your ground will shake as the mighty wolf Fenrir gallops across the land and devours everything in his path. The seas will rise as the giant snake Jörmungandr releases its tail on the bottom of the ocean and thrashes its way to the surface. An army of giants, led by the trickster god Loki, will sail across the flooded Earth, and Heimdall, the sentry of the gods, will sound his great horn to let the gods know the final battle has begun.

Amid the fighting, Odin and his legion of warriors from Valhalla will be slain by Fenrir, who will in turn be slain by Odin's son, Vidar. Heimdall and Loki will kill each other. Thor and Jörmungandr will kill each other. The world—maybe every world, for there are nine in Norse mythology[1]—will sink to the bottom of the ocean, leaving nothing behind except the void that was present at the time of creation. All will perish, save for two humans who manage to stow away in a forest, or perhaps in the roots of the sacred tree Yggdrasil, depending on the storyteller. Their survival provides vital hope that the world can, and probably will, be reborn.

I have something that borders on respect for the Norse and the way their stories map so nicely onto modern scientific understanding. Surely even you can see elements of your scientists' Big Bounce hypothesis in the Ragnarok myth.

The most enduring of your human doomsday stories come from the family of Abrahamic religions: Christianity, Islam, Judaism, and others, whose combined followers comprise more than half of your entire planet's population. The stories of each Abrahamic branch have some small discrepancies, but it's nevertheless clear that they're related.

Some of you humans will bristle at the idea that your religions are also mythologies. But the fact that they've persisted makes them no less mythical, it merely indicates they have longer-lasting support.

In these stories, The End is usually understood to be the inevitable conclusion of God's fight against the ever-advancing tide of evil. One Christian version comes to us from John of Patmos (at least, that's what modern human researchers call him, after the Greek isle where the text was purportedly written), who lived less than a century after Jesus died and whose work, the Book of Revelation, appears in the second installment of the Christian bible, the New Testament. In the book, John describes how Jesus appeared to him and told him of the impending end of the world, and how he then traveled to heaven to hear the particulars of it from a group of loose-lipped angels.

They foretold that four horsemen sent by God would herald the end of the world, bringing conquest, war, famine, and death upon it. The sinister equestrians (and their equally menacing horses—probably bitter about the whole saddle thing) wouldn't be enough to completely rid the world of evil, so God would also send seven angels to blight the land. The angels bring pestilence to the nonbelievers, turn the seas and rivers into blood, cast the world into darkness, and trigger all kinds of natural disasters. After all, it isn't really the end of the world without floods and earthquakes and fiery rain, right? I suppose the story has a happy ending for John's readers, though, for God's victory isn't complete until the world has been destroyed, all the souls judged, and the faithful resurrected into a new world. Real light-hearted stuff.

It's interesting that the Norse and Christian doomsday stories, as well as dozens of others from around your world, start with ages of violence and corruption among humans. The apocalypse is a punishment, or maybe, according to some, a gift from God. It is a final, triumphant cleansing of evil from the world. At least that's what the prophets wanted their audience to believe. From where I was standing—which was and will continue to be all around you—it looked like the stories were a means of maintaining social order.

"Everyone better be good, or I'll send another flood to wipe everything out!" said Yahweh, Enlil, Vishnu, and all the other gods who were said to decimate civilizations with great deluges.

There were a lot of floods, but regardless of the destruction method, doomsday stories served a different purpose than myths of the afterlife. They went beyond teaching about the consequences of *individual* actions, though that lesson certainly isn't absent when it's the righteous and faithful who survive in the next world. Doomsday myths are about the consequences of human behavior *as a whole*.

When the Book of Revelation was written, many believers thought the apocalypse was imminent. Who could blame them? Christians in the first century after Jesus's death were persecuted by Romans. John even wrote the book while in exile! The Christian world was at war, a war that destroyed the temple in their—and other Abrahamic religions'—holiest city. The volcano you've named Mount Vesuvius erupted violently, destroying multiple metropolises. But the world obviously didn't end, and Christians merely adjusted their timetable.

Humans have never been shy about crying apocalypse. Curmudgeonly Assyrians living five thousand years ago complained that the loss of manners and morals among the citizenry would bring about the end of the world. Johannes Stöffler, a German astrologer and mathematician, was just one of many scientists whose bad math and sloppy observations led to recklessly inaccurate apocalypse predictions, many of which falsely accused *me*. Stöffler predicted that an alignment of the planets[2] would cause a catastrophic flood in 1524, then 1528, and lived to see himself be wrong twice.

The world famously did not end at the conclusion of the Mayan long-count calendar in 2012, though the Maya never intended for that to be interpreted as a doomsday, merely the beginning of the next cycle. Be honest with me, human. Were you among the roughly 10 percent of your kind who believed the world would end then? Never mind, the past is the past. Hopefully you'd know better by this point in the book, anyway.

Even now, there are humans who believe that the pervasive diseases, intensifying storms, fires, floods, droughts, and earthquakes are yet again signs that the apocalypse is nigh. I can assure you that this is no burgeoning apocalypse, for that is technically a term that means a revelation of knowledge from God. And the blame for these disasters falls on humanity's shoulders and those shoulders alone. Your oceans catch on fire these days. That never used to happen when you relied on *me* to light your way instead of oil and coal.

If these disasters are a sign of looming doom, they harken to your species' demise, not your planet's. It takes more than moral decline and a few potion-carrying angels to destroy a whole world, especially one that I built. And I won't be terribly upset if you humans wipe yourselves out. I might be a little lonely for a few billion years, but then there will be more creatures to take your place.

But if I were a betting galaxy—which I'm not, because we

galaxies don't have any currency to gamble with—I would wager that these latest naysayers are wrong. Don't let this go to your head, but I believe you humans can learn to coexist with your planet again.

Why? Because a few hundred thousand years of spectating have taught me a lot about humanity's excellent self-preservation instincts. And because your scientists are too stubborn to give up on a question once it's been asked.

CHAPTER FIFTEEN

SECRETS

Your kind have been watching me since before you were even fully human. Two hundred thousand plus years of hunting, navigating, timekeeping, and storytelling by the light of my stars. You tracked their motion so carefully that you learned to predict in advance where they'd be. Then, in just a few centuries, you invented tools to study not only those stars, but also the planets around them, and the stars built by other galaxies. You've come so far, but even after millennia of studying my nature and telling my stories, you humans still have so much more to learn.

And since my calendar is relatively clear for the next few eons, I'm very excited to watch you figure it all out. Every year, newly trained physicists and astronomers hope they'll be the one to discover why your sun's magnetic field flips every eleven years, or what Jupiter's core is made of, or why fast radio bursts happen, or what it's like inside a black hole. Their fresh minds will join the ranks of seasoned scientists who are already working to solve the mystery of our universe's slight matter-antimatter asymmetry or observe what happens in the moments leading to a supernova explosion.

They have so many questions, and I wouldn't dare deprive them of the honor of answering them on their own, nor myself the joy of watching them stumble their way through it. But I will happily tell you how they're going about unraveling some of their most pressing me-related problems.

Before you can truly understand how I physically became the galaxy I am today, you humans are going to have to learn the secrets of dark matter. Your scientists have been dancing around the discovery of dark matter since William Thomson observed that I was heavier than I looked in the 1880s. You may know Thomson better as Lord Kelvin, the namesake of your scientists' favorite temperature scale, but he was also the first person to attribute my extra mass to what he called "dark bodies." Kelvin's contemporaries were practically clueless about the life spans of stars or the age of the universe, so he imagined these dark bodies as the cold and dead remnants of stars. The idea spread among his peers, but when a French scientist named Henri Poincaré wrote about Kelvin's work two decades later, "dark bodies" became "matière obscure," or "dark matter."

It's nearly one hundred fifty years later, and human astronomers have made the same observation over and over. In 1922, British astronomer James Jeans studied the motion of stars in my midplane, specifically their vertical velocities, and found that they were moving faster than could be explained by the mass of my visible parts. Dutch astronomer Jan Oort reached the same conclusion from similar data ten years later. In 1933, Swiss scientist Fritz Zwicky noted the mass discrepancy when he measured the velocities of galaxies orbiting in the Coma Cluster, several neighborhoods and roughly 100 megaparsecs (a megaparsec is 1,000 kpc) away. Horace Babcock measured it again in 1939 when he was sneaking a peak at Andromeda's magnificence to study galactic rotation curves. It wasn't until the 1970s, when Vera Rubin's extensive work on rotation curves provided the first reliable evidence of invisible matter influencing the motion of galaxies, that the wider community of human astronomers started to take the question of dark matter seriously.

You've seen the effects of dark matter in the orbits of stars around spirals like me, in the dispersed velocities of ellipticals and globular clusters, and in galaxy clusters whose transparent mass aids in the bending of light from background sources through gravitational lensing. But you still don't know what dark matter is made of or exactly how it's distributed throughout my body and the rest of the universe.

Your scientists have their ideas, though. Kelvin's 1884 supposition that my extra mass could be attributed to dark bodies like small black holes, brown dwarfs, and free-floating or "rogue" planets survived until the turn of your latest millennium. Your scientists called these invisible anchors "massive astronomical compact halo objects." MACHOs for short. These MACHOs are made of regular baryonic matter just like you, each just a different combination of quarks and electrons.

After you *finally* figured out how to launch projectiles into space (just barely), you learned more about the large-scale structure of the universe and how much matter it holds. Your scientists had already deduced how many atoms could have been produced in the Big Bang, and it wasn't anywhere near the number of atoms necessary to account for all the universe's mass. Thus, Kelvin's original MACHO idea was put to rest, until a small contingent of astronomers decided to investigate primordial black holes (hypothetical ones that formed right after the Big Bang) as sources of dark matter.

Still, with MACHOs mostly out of the picture, human astronomers agreed that there are two important characteristics of dark matter: it doesn't interact with the electromagnetic force, but it does interact with luminous matter gravitationally. Eventually, you noticed that it also doesn't interact with the high-energy protons and atomic nuclei that are always whizzing around our universe. Your scientists call them cosmic rays, and the fact that they can pass right through dark matter without it interrupting the strong force that holds them together indicates that dark matter doesn't interact with the strong nuclear force, either.

The only force left (that your scientists know of) is the weak nuclear force, the one that helps particles decay. It and gravity are the two less powerful forces.

Since the 1980s, some of your astronomers have considered that dark matter may be large clouds of particles that are up to a thousand times heavier than protons and interact only with forces as weak as or weaker than the weak force. They're unsure of what these particles are, but they've given them a name (possibly reclaiming a childhood barb): WIMPs. Weakly interacting massive particles. Your scientists have been looking for them for decades to no avail. They've tried detecting WIMPs indirectly by searching for the extra gamma rays produced by decaying photons as they pass through dark matter in distant galaxies, or neutrinos produced by dark matter particles interacting with photons in your sun. They've built detectors that they hoped would be sensitive enough to perceive the tiny amount of energy produced when a hypothetical dark matter particle collides with the nucleus of a normal atom, usually xenon or germanium. Your scientists have even attempted to make their own WIMPs in that big Swiss ring they call the Large Hadron Collider. It's been forty years—a long time for you humans!—and your scientists still haven't given up hope of finding the WIMPs. But their lack of success has urged them to consider other possible explanations for dark matter.

Most of the particles I've mentioned thus far—photons, Higgs boson, quarks, but not the hypothetical WIMP—are part of your scientists' standard model of particle physics. This model explains almost everything humans have ever observed about our universe, but there's always been a *problem* with how it describes neutrinos—a strong charge-parity (CP) problem, to be exact. Apparently, they're too symmetrical, too similar to their antineutrino counterparts to fit into the assumptions your physicists have made about how the universe works.

In 1977, two physicists working at Stanford University proposed solving the CP problem by adding a new rule of symmetry to the standard model. Two other American physicists separately realized that if this symmetry were spontaneously broken, it should produce a particle much, much less massive than a proton (even lighter than an electron!). These days, this theoretical particle is called the axion, but I preferred when you called it the higglet. Your scientists expect that this symmetry is broken quite often, which would make axions extremely abundant. Maybe even abundant enough to account for the entirety of the universe's invisible matter. And recently, a few physicists began to suspect that the axion might be able to solve that matter-antimatter asymmetry problem, too.

Individual axions' low mass makes interacting with normal matter difficult for them, but they do interact with magnetic fields, and they can decay into photons. One experiment at the University of Washington in Seattle is using strong magnets to coax any wayward axions into decaying into photons, which the magnets can easily detect. In 2020, there were claims that the XENON1T instrument in Italy had detected axions coming from your sun. The legitimacy of the detection is still a matter of debate, but it won't affect the search for a dark matter particle either way. Solar axions would be too energetic and hot to behave like dark matter, which needs to be cold to clump as well as it does. Though, there are some human scientists who would note here that dark matter can be hotter (i.e., move faster) if the particles are less massive.

I'll be paying close attention to this field of particle physics in the coming decades. Even if you don't find the exact particles you're looking for, I'm positive you're going to discover *something* new, because I know there's more to find.

There is a small faction of physicists who theorize that dark matter isn't matter at all, and that the answer to its riddle won't be found in a particle accelerator. Instead, they reason the effects of dark matter can be explained by tweaking Isaac Newton's definition of gravity. An Israeli physicist named Mordehai Milgrom conceived the idea in 1983, when he called for a theory of modified Newtonian dynamics, or MOND. MOND advocates believe that Newtonian gravity works only in high-acceleration environments, like Earth and your solar system. Low-acceleration environments like the outer edges of galaxies operate under different gravitational rules.

As much as I respect the desire to continuously prove Einstein wrong, MOND doesn't hold up against observations of galaxies without dark matter. If it's just a matter of gravity behaving differently at larger scales, then its effects should be observed everywhere. Your astronomers have found a couple of galaxies confirmed to be abnormally devoid of dark matter. They're called NGC 1052-DF2 and NGC 1052-DF4. Your scientists say their dark matter was stolen by bigger galaxies, and you should know that we galaxies still talk about what happened to them, only in hushed whispers.

Dark matter could be any of these, or it could be sterile neutrinos, or massive particles that interact strongly with themselves but not others. Maybe dark matter is a combination of several particles, or maybe it's something your scientists haven't even thought of yet. Once you figure out what it is, you'll understand a whopping 32 percent of our universe's total matter-energy content. What's the other 68 percent? That's a question that has haunted human cosmologists for twenty years. (I've watched idly as some of them lost sleep and tore their hair out over it. Those take their jobs far too seriously.)

You may have heard that dark energy is the force that makes

the universe expand. This is a common misconception among humans, as the universe could expand even without dark energy. Most human scientists, even that dunce Einstein, agreed by the 1930s that the universe was expanding, but they assumed the expansion was decelerating. Two teams of astronomers on practically opposite sides of your planet spent years studying distant supernovae, the special kind that human astronomers use as standard candles. They hoped to learn exactly how fast the expansion was slowing down, but in 1998, both independently announced that the expansion was speeding up instead.[1] They invented a hypothetical type of energy that exerted a repulsive force in the universe, and they called it "dark" because they had no clue what it could be.

One idea they've had is that dark energy is just an inherent quality of space, as fundamental and inevitable as gravity. And just as more mass equals more gravity, more space equals more dark energy, so the strength or density of dark energy stays constant even as the universe expands. Your scientists have called this convenient quality the cosmological constant, depicted in their equations as Λ. The most widely accepted model of our universe among your scientists is the Λ-CDM model. It dictates that the cosmological constant and cold dark matter are necessary to explain your observations.

A less popular idea is that dark energy is the work of yet another quantum field, this one called "quintessence." It means the fifth essence of the universe, after normal matter, dark matter, radiation, and neutrinos (i.e., baryons, whatever dark matter is made of, photons, and leptons). Unlike the cosmological constant, quintessence is not an unavoidable consequence of having empty space. The density of quintessence changes as the universe expands, which means the strength and influence of quintessence change over time.

The best hope your scientists have of understanding dark energy is to keep measuring the universe's expansion rate at different times, and in all directions. This is where the finite speed of light works to your advantage. To look further back in time, you humans just need to figure out how to see more distant (and therefore likely much fainter) objects. Your imaging and measuring technologies have finally reached the point where that's possible. For a brief moment, I wasn't sure if it would ever happen because your earthling brain trust seemed more focused on Snuggie technology. You do know it's just a blanket with armholes, right?

Since 2013, hundreds of scientists from all over your world have been collaborating on the Dark Energy Survey (DES). Using an extremely sensitive camera mounted on one of their older telescopes in Chile, they've mapped more than three hundred million galaxies looking back billions of years. If the acceleration of the universe's expansion has always been the same—its acceleration, not its speed, mind you, human—that's a point for the cosmological constant. A variable acceleration would point towards quintessence.

The DES team still has a lot of data to comb through, but I suspect that they'll have an exciting announcement not too long after this book is made available. If not, human scientists have high hopes for the upcoming Nancy Grace Roman Space Telescope (formerly the Wide-Field InfraRed Survey Telescope, WFIRST). Named after the first chief of astronomy at your National Aeronautics and Space Administration, the Roman telescope will capture expansive images with its camera's three hundred million pixels. NASA is scheduled to launch this long-anticipated successor to the Hubble telescope into orbit around your planet in the mid 2020s, armed with instruments to study everything from planets to galaxies to the expansion of our universe.

There is one question about me that you humans ask more than any other. If I had one of those flimsy pieces of paper you call a dollar for every time one of you pondered this question, I'd be richer than any human. It's been asked so often that most of you have already taken a side in the matter, probably because billions of dollars' worth of movies, TV shows, video games, and books shove a particular answer to this question in your face often. You want to know: Is there other life out there?

In the supposedly grand scheme of human history, this is a relatively new question for humans to be asking. It wasn't until one of you used a telescope to look at Mars for the first time in 1609 that you realized there were other planets similar to your Earth. After that, it was only a brief time before you started to wonder if those other rocky planets could hold life, too. In 1877, an astronomer named Giovanni Schiaparelli described a network of long straight lines on the surface of Mars that he called "canali" in his native Italian tongue. He meant "channels," but yet again, because you humans never managed to institute a single global language, it was interpreted as "canals" when Schiaparelli's work was translated to English. For the rest of the nineteenth century, many human astronomers, including an especially stubborn man named Percival Lowell, took these canals to be evidence that life once existed on Mars.

But there were never any canals. And the channels that Schiaparelli described were in fact a trick of the faulty, overactive human brain. He saw a blurry image through his aggressively rudimentary telescope, and his mind added straight lines where there were none. How do you ever trust anything you see with that soggy, lying brain of yours?

Canals or no, human astronomers continued to ask whether the other planets in your solar system could support life, a question they're *still* trying to answer. Before long, they started to speculate about planets around other stars, which your astronomers now call exoplanets.

It must be so hard to be a human, with your inconvenient corporeal form and no power to control what wavelength your weak eyes can see. If you could be everywhere at once like me, you would just *know* about all the planets. If you had keener eyes capable of ignoring starlight, you'd be able to just *see* them. Instead, your astronomers have had to come up with creative and mostly indirect methods of finding these exoplanets.

Or should I be calling them endoplanets since they're in me? I think I'll simply call them planets to avoid confusion. Your solar system isn't special to anyone but you, but it does tickle me to see you work so hard to understand it.

These creative detection methods aren't the most efficient. Out of my more than one hundred billion planets, they've found only about five thousand since the first was discovered in 1992.[2] By far the most successful method is what your astronomers call transit photometry. When a planet passes directly between you and the star it orbits, your astronomers can measure the amount of starlight the planet blocks. The dip in brightness is proportional to the planet's size, and if you wait long enough to see the planet transit more than once, you know how long the planet's period is. Once you know the period, it really is just a matter of plugging it and the host star's mass into a formula developed four centuries ago to find the distance between the planet and its star, and then it's just a few more equations and some simplifying assumptions before you know the planet's temperature. Who could have guessed there would be so much information in a planet's shadow?[3]

It's a good thing your astronomers did, because about 75 percent of the planets they've discovered were found using the transit method, even though most of my planets (more than 99 percent!) don't have the "right" orientation to transit from your perspective. They found another roughly 20 percent by measuring the motion of host stars as they're tugged by the gravity of their planets. More specifically, human astronomers measure the velocity of stars as they move towards and away from Earth, which your scientists call radial velocity. The faster the radial motion, the stronger the planet's tug on its star, and therefore the more massive the planet.

At first, the focus was on finding as many planets as possible. Each planet added to your understanding of the planetary population. You found that planets are common around main sequence stars and witnessed the amazing diversity of planets I've made (with a little help from my stars, of course). You even found those hot Jupiters I created once by mistake and then kept making because they were amusing.

But if you want to know whether or not there's life out there, it's not enough to know that there are planets. You must know what it would feel like to stand on that planet's surface, or swim or float, as the case may be.

Your scientists have a lot of work to do before they can imagine the surface of any planet that vividly (except for Mars), but they do have some crude measures of habitability. All of them are, of course, rooted firmly in classic anthropocentricism. Oh, most life on your planet depends on water? No duh, 70 percent of your planet is covered in the stuff! That doesn't mean that life elsewhere will depend on water, too.

Though I do see how you would assume as much. Water is exceptionally good at dissolving matter into its smaller parts. That makes it considerably easier to build those parts up into something that can one day question what it means to be alive.

With water on the brain, human astronomers in the 1950s defined the circumstellar habitable zone, sometimes called the Goldilocks zone, as the range of distances from a star where a planet could hold on to liquid water at its surface. Any closer and the heat from the planet's host star would evaporate the water away. Any farther and the water would freeze. Anyone who's ever lived on a planet—plus me, who's seen enough planets from the outside—knows that surface temperature also depends on the planet's atmosphere, how reflective the surface is, and what's going on inside. Unsurprisingly, these factors often get left out of habitability calculations, yet planets still get classified according to whether they're in their star's habitable zone.

Trust me when I say that humans have never found irrefutable evidence of extraterrestrial life. Your astronomers are not hiding aliens from you.[4] I don't think they could, even if they wanted to. Your world is too connected now, and as a group, your astronomers are awfully bad at keeping secrets—beneath their lab coats and scientific jargon, they're relentless gossips. They don't know if there are aliens, but there are a few human astronomers who have taken the concept of the Goldilocks zone and extended it to a *galactic* habitable zone. They want to know where in my body the aliens *would* be. Humanlike aliens, that is.

Those few astronomers would say that in an astronomical context, human life needs three things to survive: metals, protection from radiation, and time. It seems like a rather simplistic list to me—I've heard quite a few of your kind say they'd *literally* die without their morning cup of coffee—but I'm a big enough galaxy to admit that human needs are probably the one thing you know more about than I do.

The metals you need are created in my stars, which means you'll find more of the carbons, nitrogens, and oxygens that you need in star-dense areas. Most of my stars are concentrated near my center, with fewer as you move towards my edge. Indeed, human astronomers have noticed a downward metallicity trend with increasing radius, though there are some exceptions to consider. Those little galaxies that I eat can be rich in metals, which are spread around my halo and outer disk as I rip them apart. And if I want, I can also push some metals around using the winds given off by Sarge and my stars. But for the most part, if you want a lot of heavy elements, you should be looking towards my core.

This requires a delicate balancing act with your second requirement: protection from radiation. Specifically, it's the high-energy UV, X, and gamma radiation, the type produced by supernova explosions and, to a lesser degree, stars minding their own business. Your delicate human bodies can't handle it, even the supposedly strong ones—slurping protein shakes and lifting a few measly hundred pounds as if it will save them from total annihilation. But supernovae aren't the only source of dangerous radiation, just the most powerful. You could also be fried by a particularly strong gamma-ray burst, active galactic nuclei, and the millions of high-energy cosmic rays that can pass right through your body without you knowing it.

So humans need to be where the stars are so that you can use the metals they produce, but you also need to be where the stars aren't so your cells don't start degrading or rapidly mutating. It's already a tough needle to thread, but of course there's more.

Human life needs *time*. You evolutionary prima donnas need billions of years in a stable environment to develop, which means you wouldn't be caught alive around one of my shorter-lived O or B stars. It also means you can't handle any friendly stellar flybys that could alter Earth's orbit or rip you away from your precious sun. That's right, your sun isn't allowed to have any friends visit because it would probably kill you. And you thought your parents had strict rules! This eliminates my bulge, where most stars get to pass by their pals at least once every billion years.

What do all these seemingly contradictory restrictions mean

for the fabled galactic habitable zone? Well, according to human astronomers, who are *totally* unbiased, the aliens are likely to be found in a ring between 7 and 9 kpc from my core. Sound familiar? Well, it should, because your solar system is smack-dab in the middle of it. It's mighty convenient that the aliens your scientists are looking for should be close to you (or on the complete opposite side of my disk), because most of the planets they've found are within a kiloparsec of your solar system.

Once your astronomers successfully identify a planet that meets all of their restrictive habitability requirements, there are a couple of different ways they can try to figure out whether it's indeed inhabited.

The first is to look for the by-products of life, which they call biosignatures. Most of the biosignatures your astronomers look for are gases that living things produce in their bodies and then release into the atmosphere. The phosphine that some of them recently claimed to have found on Venus—then unclaimed, and then claimed again—is a biosignature because it's strongly associated with your wasteful biological processes, though it can be produced in small quantities in other ways. There's always the possibility of false positives when it comes to biosignatures. Oxygen and methane are others that you've probably heard of.

There are other, less talked-about biosignatures, like light reflecting off living creatures (plants, algae in oceans, etc.) in a specific way. It's also possible to detect seasonal patterns in certain gases, like CO_2 abundances waning and waxing as photosynthesizing creatures grow and die. To find these and other gaseous biosignatures, human astronomers came up with two methods rather confusingly named transit spectroscopy and transmission spectroscopy.

Remember that transit photometry measures the changing brightness as a planet passes in front of its star. Transit spectroscopy, on the other hand, measures the planet's transit depth at different wavelengths to learn what its atmosphere is made of. Depending on the atmosphere's composition, it will be more opaque at some wavelengths, and more transparent at others. The opacity of the planet's atmosphere affects how much starlight it blocks, and therefore the transit depth.

With transmission spectroscopy, astronomers measure the changing spectrum of the star's light as it passes through its transiting planet's atmosphere. Some of the star's photons are blocked and absorbed by molecules in the atmosphere, so your astronomers see a gap in the spectrum at those photons' corresponding wavelengths of light.

Thus far, both methods have worked only for planets with thick atmospheres like Jupiter, but human astronomers are confident that the upcoming James Webb Space Telescope (JWST) will change that. The telescope was previously called the Next Generation Space Telescope, and there was a push by some astronomers for the name to be changed again so as not to memorialize a man who discriminated against and persecuted his colleagues simply for daring to love another human with the same-shaped fleshy bits.[5] Sounds like a reasonable request to me; that kind of silly, small-minded human thinking shouldn't be celebrated.

With a collecting area more than six times bigger than the Hubble spacecraft's, JWST (or whatever it ends up being called) claims it will be able to see the atmospheres of small rocky planets. It should also be able to see the first stars and galaxies forming and peer through dust clouds to see new stars and planets form.

Development for JWST began in 1996 with a plan to launch in 2007. Unexpected delays and money issues pushed the launch back to 2010, then 2013, 2018, 2019, 2020, and finally 2021. Human astronomers joked that it would never happen, and some still have nightmares about the telescope not unfolding properly when it does eventually reach orbit. But the telescope did launch successfully in 2021, on Christmas day no less! Astronomers and casual space enthusiasts around the world lauded the launch as the best gift they could have hoped for.

The other way human astronomers search for alien life is by hunting for what they call technosignatures. Assuming that life-forms on other planets evolve the same needs as you and take a similar technological path once they become intelligent, your astronomers believe they could find evidence of alien technologies.

Really, technosignature hunters are looking for proof of intentional manipulation of their environment. That could appear as chemical or light pollution, giant constructed objects like a Dyson sphere built to collect maximum stellar energy, or coherent electromagnetic signals. Your astronomers have been searching for encoded radio signals from other star systems since the 1960s. An American radio astronomer named Frank Drake originally spearheaded the work until the institute for the Search for Extraterrestrial Intelligence eventually took it up. Founded by Jill Tarter[6]—one of the few humans for whom I have nothing but respect—SETI's work has been ridiculed by many human astronomers, but it's always been one of my favorite human organizations because it dared to wonder what other interesting creatures I might hold.

Clearly, your astronomers are trying to answer the question of extraterrestrial life. But I think that when some of you ask this question, what you really want to know is, are there interstellar alien networks? You want to know if your Star Treks and Star Wars and Guardians of the Galaxies could be real. You reeeally just want to know if faster-than-light travel is possible, don't you?

Well, that's for me to know, and hopefully for you and your scientists to find out. Just as you humans are on your own to figure out how black holes got supermassive so quickly. Or where all the intermediate mass (100–100,000 solar masses) black holes are. Or whether there's a secret ninth planet in your solar system. Or why the IMF is what it is. Of course, I've already told you it's because I don't like making stars I know will die soon, but who is listening to me? A better question is whether or not the IMF is universal. Do all galaxies treat their stars the same?

These are just some of the questions your scientists are trying to answer when they're not teaching, attending talks by their peers, or begging for financial support from any—and I mean *any*—source of funding that will listen. Given their desperation, I have no doubt they'll figure it all out any moment now. For me, though, the real breakthrough will be when your kind start asking the questions no one has even *thought* of yet.

If you, tiny reader, are not going to join these scientists, be

patient as they stumble their way towards understanding. They are, after all, only human. All I can do is wish your scientists luck and cross my figurative fingers that they put on a good show. If only I had some popcorn.

And when your astronomers learn something new about me or one of my galaxy companions, I hope you are just as thrilled as they are about the discovery. Especially now that we've become so much better acquainted. From your stunted perspective, it may take a long time to get there, so for now you should really learn how to live *with* your planet instead of just on it. I assure you, you aren't ready to come face-to-face with the rest of me...yet. But, if you somehow wind up, unharmed, in the far reaches of my glorious body, maybe I'll mention you in my next note to Andromeda.

I have worked so hard to hold my metaphorical tongue in this book and tell you only what the observant humans had already learned, but there is one secret that I'll give you and your scientists freely, one that I've only ever told one other galaxy. After wallowing in my own self-pity for billions of years, my journey to accepting my own awesomeness made me realize my true passion: inspiring others. Stars, galaxies, even hairy meatbags like you; I want to light a fire—literal or figurative—in them all!

This whole autobiography has been about inspiring you to *do something*. Ask questions about the world around you and find real answers to them. Decide you deserve better and fight to clean your sky of all kinds of pollution—trust me, I'm worth it. Or make beautiful art that people can talk about long after you've shuffled off your particular mortal coil. Here's a hint: timeless art captures timeless subjects, and nothing in your puny human life will stick around longer than me.

We galaxies have a saying. Roughly translated, it's "You can lead a fleshworm to the stars, but you cannot make it wonder." But I have yet to come across something that I cannot do, so ad astra, human. May my stars guide your way to a wondrous future full of stories. I'll be listening.

ACKNOWLEDGMENTS

I've wanted to write a book for as long as I can remember, but for a long time, I didn't think I could. So, my first thanks has to go to Jackie Slogan, the teacher who always believed I'd be able to do whatever I set my mind to. My second goes to my mom for making me fall so in love with reading books that I dreamed of writing one.

Thank you to my agent, Jeff Shreve, who sent the best cold email I've ever received when he asked me if I'd ever thought of writing a nonfiction book. Thank you again, Jeff, for not laughing when I told you I wanted to write a book from the galaxy's perspective. And thank you to Matthew Stanley for giving Jeff my name.

A whole bouquet of thank-yous to my fabulous editor, Maddie Caldwell, who was always firmly honest with me when something wasn't working, and enthusiastically honest when something was. Maddie, this book would have been so, so much worse without your help. And thank you to Jacqui Young for the keen eye and the top-notch jokes.

Thank you to my oldest friend, AnnaMarie Salai, for going with it when I asked her to draw a sentient galaxy with a personality. Thank you to everyone at Grand Central Publishing who worked on this book, from the cover to the proofreading, printing to marketing. I haven't met you all, but I'm so grateful for the part you played in making this dream of mine a reality.

Thank you to everyone who read through and fact-checked chapters, even though they must have seemed so strange without the context of the rest of the book. To future Dr. Luna Zagorac and current doctors David Helfand, Kathryn Johnston, Dreia Carrillo, Emily Sandford, Abbie Stevens, Jorge Moreno, and Kartik Sheth: your notes were invaluable! Thank you to Steve Case and David Kipping for checking me on the history and science.

A huge thank-you to my partner, William, who read chapters, helped me work through writer's block, and put up with my mood swings as I got into character—especially during the broodier chapters. Thank you again for keeping Kosmo from pouncing on me when I was in my writing fort.

He can't read this, but thank you to Kosmo, the fuzzy love of my life, for keeping me company as I wrote this book. When my mom read an early draft, she said the Milky Way sounded like a cat. I told her that was fair, because whenever I needed to get into the Milky Way's voice, I would look at Kosmo and the way he blinked indifferently at me even though he relied so completely on my care, and I would think, "Yup, that's probably how a nearly omniscient galaxy would look at me, too." So thank you for being my inspiration, Kosmo. I guess you have more than one job after all.

And finally, thank you to *you*, my dear readers. Thank you for listening to what the Milky Way and I have to say.

NOTES

A note on these notes: they are not a bibliography or a works cited list. If you're interested in that, a dynamic list of the academic papers I consulted while writing this book can be found on my personal website. Instead, these notes are full of extra tidbits of information that I thought you might enjoy, along with some extra reading about things too mundane for the Milky Way to bother mentioning. There is even some astronomy community *behind the scenes* gossip in these notes, because the galaxy was right when it said we can't keep our mouths shut.

CHAPTER ONE: I AM THE MILKY WAY

1 We humans don't have much need for numbers bigger than a trillion in our everyday lives, but that hasn't stopped us from coming up with words for really big numbers anyway. For example, the word for $10^{10,000}$ is ten tremilliatrecendotrigintillion. You can find other equally tongue-twisting number words at **Landon Curt Noll, "English Names of the First 10000 Powers of 10 - American System Without Dashes," Landon Curt Noll (blog), https://lcn2.github.io/mersenne-english-name/tenpower/tenpower.html.**

2 These hypothetical free-floating brains were called Boltzmann Brains, named after Ludwig Boltzmann, who did NOT come up with the idea. Most scientists would dismiss Boltzmann Brains as silly, but that doesn't stop physicists from getting into the most frustrating conversations about whether all of human existence is just one random brain floating for an instant in the universe.

3 Brown dwarfs are the limbo space between planets and stars. They aren't massive

enough to kick-start and sustain hydrogen fusion in their cores, though some of them get massive enough to fuse deuterium (also called heavy hydrogen) for a short while. Astronomers sometimes joke that brown dwarfs are failed stars, but we're still trying to figure out where the mass cutoff between success and failure is. A great research group called BDNYC based at the American Museum of Natural History is focused on better understanding brown dwarfs.

4 Hot Jupiters are massive planets (more than one hundred times the size of Earth) that orbit close enough to their stars to orbit them in a few *days*, instead of the twelve years it takes our Jupiter to orbit the sun. Once we discovered one in 1995, astronomers were puzzled about how a planet that big could be so close to its star. Did it form far from the star and migrate in, or did it form there? It turns out that both are possible, under certain conditions!

5 Ancient Egyptians believed the Nile flooded every year because the goddess Isis wept for her husband Osiris. When the star we call Sirius was visible at sunrise, they knew the tears were soon to come and nourish their fields for the next season's crops. **David Dickinson, "The Astronomy of the Dog Days of Summer," *Universe Today*, August 2, 2013, https://www.universetoday.com/103894/the-astronomy-of-the-dog-days-of-summer/.**

6 The International Dark-Sky Association is a nonprofit that tracks and fights the effects of light pollution. Most humans alive today have an obstructed view of the night sky, but the DSA provides advice for how you can help change that. **"Light Pollution," International Dark-Sky Association, February 14, 2017, https://www.darksky.org/light-pollution/.**

7 When I give public talks, people often ask me why we should study space at all. Aside from the fact that pursuing knowledge for the sake of learning is a noble human act, astronomy research has provided so many practical benefits to society. **Marissa Rosenberg, Pedro Russo, Georgia Bladon, and Lars Lindberg Christensen, "Astronomy in Everyday Life," *Communicating Astronomy with the Public Journal* 14 (January 2014): 30–35, https://www.capjournal.org/issues/14/14_30.pdf.**

CHAPTER TWO: MY NAMES

1 Dung beetles can't make out individual stars, but they can see the entire stream of the Milky Way stretch across the sky, and they use it to orient themselves as they roll their balls of poop backward towards their homes. Some migratory birds, like the indigo bunting, use the north star Polaris to guide them as they fly. For more examples, see **Joshua Sokol, "What Animals See in the Stars, and What They Stand to Lose," *New York Times*, July 29, 2021, https://www.nytimes.com/2021/07/29/science/animals-starlight-navigation-dacke.html.**

CHAPTER THREE: EARLY YEARS

1 Not even the Milky Way can resist Julie Andrews's charm or deny the cinematic merit of *The Sound of Music*.

2 Probably the most well-known and widely used of these simulations comes from the Illustris project. You can learn more about it at **Illustris, https://www.illustris-project.org/.**

3 Not all elements are formed in the cores of stars. Heavier elements like silver and gold are formed in energetic events like neutron star collisions. For an overview of formation mechanisms for different elements, see **Jennifer A. Johnson, Brian D. Fields, and Todd A. Thompson, "The Origin of the Elements: A Century of Progress,"** ***Philosophical Transactions of the Royal Society A: Mathematical, Physical and Engineering Sciences*** **378, no. 2180 (September 18, 2020): 20190301, https://doi.org/10.1098/rsta.2019.0301.**

4 The universe as a whole is cooling down, but interactions between gas particles in coalescing galaxy clusters actually cause the gas to heat up. **Matt Williams, "The Average Temperature of the Universe Has Been Getting Hotter and Hotter,"** ***Universe Today,*** **November 14, 2020, https://www.universetoday.com/148794/the-average-temperature-of-the-universe-has-been-getting-hotter-and-hotter/.**

5 When thinking about technological and scientific advancement, it's easy to assume that all societies should follow the same path and make the same stops. But tools develop in parallel with a society's needs, and not every group needs a way to count or distinguish between large numbers. That doesn't automatically make them less advanced. **Caleb Everett, "'Anumeric' People: What Happens When a Language Has No Words for Numbers?,"** ***The Conversation,*** **April 25, 2017, https://theconversation.com/anumeric-people-what-happens-when-a-language-has-no-words-for-numbers-75828.**

6 Many cultures around the world (Greek, Mesopotamian, Egyptian, etc.) believed that the sky was the home of the gods and that its behavior reflected their will. The various peoples of southern Africa tended to see the sky as a solid dome that separated our world from something... else. Stars were either pinpricks poked in that dome, or lights hanging from it on strings. For an overview of night sky beliefs from around the world, see **"African Ethnoastronomy," Astronomical Society of Southern Africa https://assa.saao.ac.za/astronomy-in-south-africa/ethnoastronomy/.**

7 Mayflies do live impressively short lives, but it's not entirely accurate to say they live for a day. They can stay in their aquatic larval stage for months up to a few years. When they emerge from the water with their wings, males can live a

couple of days, while female mayflies can live for as little as FIVE MINUTES, just enough time to mate and lay their eggs.

8 These same gravitational interactions slowing our rotation are also causing the moon to move away from us by about 1.5 inches every year. Eventually, the moon will be so far away that it looks smaller than the sun in the sky, and total solar eclipses won't be possible anymore. But that won't happen for hundreds of millions of years.

9 I took a class on stellar aging techniques in graduate school, and we basically spent the whole semester working our way through a single review of methods, here: **David R. Soderblom, "The Ages of Stars,"** ***Annual Review of Astronomy and Astrophysics*** **48, no. 1 (August 2010): 581–629, https://doi.org/10.1146/annurev-astro-081309-130806.**

10 Several papers have pointed out that there's a sweet spot in stellar age for hosting life, and our sun is pretty much in the sweetest point on the timeline. **Abraham Loeb, Rafael A. Batista, and David Sloan, "Relative Likelihood for Life as a Function of Cosmic Time,"** ***Journal of Cosmology and Astroparticle Physics*** **8 (August 18, 2016): 040, https://doi.org/10.1088/1475-7516/2016/08/040.** Though one of the authors of that paper has recently tarnished his reputation by forcefully pushing the idea that an asteroid found in our solar system a few years ago was sent by aliens to study our solar system.

CHAPTER FOUR: CREATION

1 More than 99 percent of the four billion species that scientists think have evolved on Earth based on fossil record have gone extinct. Our planet has gone through multiple mass-extinction events. **Hannah Ritchie and Max Roser, "Extinctions," Biodiversity, Our World in Data, 2021, https://ourworldindata.org/extinctions.**

CHAPTER FIVE: HOMETOWN

1 Astronomers have known for a while that the Milky Way's disk is warped, and thanks to the Gaia spacecraft, they recently determined that the warp was caused by interactions with a satellite galaxy. **E. Poggio, R. Drimmel, R. Andrae, C. A. L. Bailer-Jones, M. Fouesneau, M. G. Lattanzi, R. L. Smart, and A. Spagna, "Evidence of a Dynamically Evolving Galactic Warp,"** ***Nature Astronomy*** **4, no. 6 (March 2, 2020): 590–96, https://doi.org/10.1038/s41550-020-1017-3.**

2 One of my advisors and several grad students in my department worked on characterizing the orbit of the Sagittarius stream to study its origins. To learn

more about the stream and its star formation history, read **Nora Shipp, "Galactic Archaeology of the Sagittarius Stream,"** ***Astrobites*****, June 20, 2017, https://astrobites.org/2017/06/20/galactic-archaeology-of-the-sagittarius-stream/.**

3 We don't have one in English, but German is the language where you can find the best words describing almost any complicated concept. In this situation, a Deutschlander might use the word "Notnagel" to refer to a companion of last resort.

4 For a less biased description of the Triangulum galaxy, look to NASA's archive of galaxies. **Rob Garner, ed., "Messier 33 (The Triangulum Galaxy)," NASA, February 20, 2019, https://www.nasa.gov/feature/goddard/2019/messier-33-the-triangulum-galaxy.**

5 Magellanic spiral galaxies may be kind of common in the universe, but they are relatively rare close to massive galaxies like the Milky Way. **Eric M. Wilcots, "Magellanic Type Galaxies Throughout the Universe," in "The Magellanic System: Stars, Gas, and Galaxies," ed. Jacco Th. van Loon and Joana M. Oliveira,** ***Proceedings of the International Astronomical Union*** **4, no. S256 (July 2008): 461–72, https://doi.org/10.1017/s1743921308028871.**

6 Henrietta Swan Leavitt was just one of at least eighty women employed by Edward Pickering between 1877 and 1919. These brilliant women analyzed vast amounts of stellar data but were still disrespected by many contemporaries who called them "Pickering's Harem."

7 These isolated void galaxies are rare, but they form in an interesting way. Their fate, though, is pretty similar to a typical galaxy's. **Ethan Siegel, "What Is the Ultimate Fate of the Loneliest Galaxy in the Universe?,"** ***Forbes*****, December 18, 2019, https://www.forbes.com/sites/startswithabang/2019/12/18/what-is-the-ultimate-fate-of-the-loneliest-galaxy-in-the-universe/?sh=d479b0c566a2.**

8 These disks were all thirty inches across and a few millimeters thick, drilled with holes to hold the fibers that carried the observed target's light to a spectrograph. Any member of the SDSS collaboration can request to own an old disk, and they often do creative things with them. One of the faculty members in my graduate department turned his old disk into a table. **SDSS-Consortium, "Serving Up the Universe on a Plate," Max Planck Institute for Astronomy, July 14, 2021, http://www.mpia.de/5718911/2021_07_SDSS_E.**

CHAPTER SIX: BODY

1 A common misconception is that the Milky Way's gravity is dominated by its supermassive black hole. While Sgr A* might be the heaviest single object in

the galaxy, the combined mass of all the other stars in the bulge is roughly ten thousand times greater than the black hole's.

2 The Milky Way knows this because it can feel all of its stars, but humans know about these bulge interactions thanks to me! This result came out of one of my original research projects in graduate school. **Moiya A. S. McTier, David M. Kipping, and Kathryn Johnston, "8 in 10 Stars in the Milky Way Bulge Experience Stellar Encounters Within 1000 AU in a Gigayear."** ***Monthly Notices of the Royal Astronomical Society*** **495, no. 2 (June 2020): 2105–11, https://doi.org/10.1093/mnras/staa1232.**

3 FYI, the Milky Way's "eyes" in the illustrations are supposed to be globular clusters.

4 Astronomers were perplexed by a dark matter–less galaxy a few years ago, and the mystery has only kind of been solved. **Ethan Siegel, "At Last: Galaxy Without Dark Matter Confirmed, Explained with New Hubble Data,"** ***Forbes*****, June 22, 2021, https://www.forbes.com/sites/startswithabang/2021/06/22/at-last-galaxy-without-dark-matter-confirmed-explained-with-new-hubble-data/?sh=7b8a6edb63dc.**

5 To learn more about Vera Rubin's life and amazing contributions to science, see **Tim Childers, "Vera Rubin: The Astronomer Who Brought Dark Matter to Light," Space.com, June 11, 2019, https://www.space.com/vera-rubin.html**.

6 This isn't just a weird figure of speech on the Milky Way's part. Caroline Herschel actually spoon-fed her brother while he read and worked on his telescopes. They wrote about it in journals, and images of her feeding him have appeared in museums.

7 In case you were curious, the man who coined the term "parsec" is NOT the same Dyson who came up with the idea of the Dyson sphere, an artificial object built to capture maximum solar energy. That would be Freeman Dyson. Nor is he the same as James Dyson, who invented some nice vacuum cleaners.

8 My grad school advisor and another member of our research group published the first reliable discovery of an "exomoon," a moon orbiting a planet outside of our solar system! The Hubble Space Telescope's observation schedule is public information, but Alex and David (the authors of the paper) didn't realize that. It was a big news story in the astronomy world, so they got a lot of calls from science journalists asking for interviews before they were ready to talk about it. It pushed them to work on the data faster. **Alex Teachey and David M. Kipping, "Evidence for a Large Exomoon Orbiting Kepler-1625B,"** ***Science Advances*** **4, no. 10 (October 3, 2018): eaav178, https://doi.org/10.1126/sciadv.aav1784.**

CHAPTER SEVEN: MODERN MYTHS

1 My personal thoughts on astrology are a bit more complicated than the Milky Way's. I know people who use it as a casual hobby or as a way to gently guide their decisions. As long as someone doesn't use astrology to hurt someone else, I won't begrudge them its use. But there are parts of the world (especially in South Asian countries) where astrology is used in a more discriminatory way. So I can't make a blanket statement that astrology is harmless.

2 This RA and Dec system is the most commonly used, probably because it's more convenient than the others, especially for naked-eye objects. Whatever latitude you're standing at...that's the declination of the star right above your head. But as our solar system orbits the Milky Way and as our Earth precesses, or wobbles on its axis, the grid moves with us, and individual objects' coordinates change. Astronomers make up for this by including a reference date, called an epoch, that tells how the coordinate grid aligns with the stars.

CHAPTER EIGHT: GROWING PAINS

1 Spanish (and similar Latin-language) speakers will see the connections among names of days, planets, and Roman gods more easily: lunes/Luna/Moon, martes/Mars, miércoles/Mercury, jueves/Jupiter, and viernes/Venus. Whereas English and other Germanic languages get their day names from Norse mythology: Tuesday is for the war god Tyr; Wednesday is for the All-father Odin, or Woden in its less anglicized form; Thursday is for thunderous Thor; and Friday is for the lovely Frigg.

2 Astronomers have gone back and forth on whether most stars are born alone, but the current consensus seems to be that stars are most commonly born in pairs, or even bigger groups. See **Scott Alan Johnston, "Our Part of the Galaxy Is Packed with Binary Stars,"** ***Universe Today*****, February 24, 2021, https://www.universetoday.com/150274/our-part-of-the-galaxy-is-packed-with-binary-stars/.** For more on stellar multiplicity, see **Gaspard Duchêne and Adam Kraus, "Stellar Multiplicity,"** ***Annual Review of Astronomy and Astrophysics*** **51, no. 1 (August 2013): 269–310, https://doi.org/10.1146/annurev-astro-081710-102602.**

3 We call M stars red dwarfs and O stars blue giants because their energy spectra do peak in the red and blue, respectively. The color of a star depends on its temperature. Wien's displacement law says that the hotter a star is, the smaller the wavelength of light where its peak emission will be. Because O stars are hotter, they emit most of their light at short wavelengths, which look blue to our human eyes.

4 The Milky Way is just making fun of us astronomers. We don't actually get that passionate about the initial mass function. Over time, scientists like Edwin Salpeter (I was friends with his grandson in college!) and Pavel Kroupa modeled slightly different functions to describe the way stellar masses are distributed. The different functions are useful for different mass ranges of stars (Kroupa deals with low-mass stars while Salpeter describes stars more massive than the sun) and different stellar environments. I did a lot of work modeling the stellar distribution in the bulge, and I usually used the Chabrier IMF because it covers a wide range of stellar masses. Also, it sounds fancy.

5 Helium atoms are bigger than hydrogen, so it takes more energy to fuse them together. Astronomers have different names for the processes that produce helium out of hydrogen. The proton-proton (or p-p) chain describes the mechanism of fusion in low-mass stars, while the C-N-O cycle is used in stars more massive than the sun where carbon is available as a catalyst. (The *C* in C-N-O is carbon; the others are nitrogen and oxygen.) Once helium has been created, the triple alpha reaction combines helium atoms to make carbon.

6 Scientists still aren't sure if the sun will engulf Earth when it expands as a red giant. There are too many factors to consider, like how much mass the sun will shed, or whether any of the inner planets' orbits become unstable. It's also possible that gravitational interactions will force Earth to break its orbit around the sun, which would be disastrous in a whole other way. **Ethan, Siegel, "Ask Ethan: Will the Earth Eventually Be Swallowed by the Sun?,"** ***Forbes*****, February 8, 2020, https://www.forbes.com/sites/startswithabang/2020/02/08/ask-ethan-will-the-earth-eventually-be-swallowed-by-the-sun/?sh=48c6f23c5cb0.**

7 In 2017, astronomers detected a signal from two neutron stars colliding, and in the aftermath of the explosion, they detected A LOT of gold and platinum. A Jupiter's worth of gold, actually. Since then, astronomers have figured out that neutron star mergers produce more gold and other "r-process elements" than supernovae and neutron star–black hole mergers. **Robert Sanders, "Astronomers Strike Cosmic Gold,"** ***Berkeley News*****, October 16, 2017, https://news.berkeley.edu/2017/10/16/astronomers-strike-cosmic-gold/.**

8 Well, I care what a neutrino is! It's a small fundamental particle, in the lepton fermion group along with electrons and taus. Neutrinos are so light that they don't interact much with other particles, and scientists haven't been able to measure their mass precisely. They're produced almost every time atoms interact, and once they're formed, they can oscillate between different "flavors" of neutrino through a mechanism that scientists don't understand. Neutrinos are interesting and mysterious, and I think the Milky Way probably says it doesn't care about them because it knows they're so cool.

CHAPTER NINE: INNER TURMOIL

1 For more on this, check out the butt-tastic "Gluteology" episode of Alie Ward's award-winning podcast *Ologies*. It features Natalia Reagan, a primatologist, anthropologist, and, apparently, butt expert.

2 Saturn's rings are the most famous in our solar system, but the other gas giants also have their own, less spectacular rings. Despite the rings' abundance in our solar system, astronomers don't know how common rings are around other planets, because it's so hard to find them! There's no reason to think our solar system is special, however, so it is likely giant exoplanets are sporting some mighty fine rings of their own.

3 After observing S2 for more than twenty years, astronomers used its orbit to confirm one of Einstein's predictions called Schwarzschild precession. **GRAVITY Collaboration: R. Abuter, A. Amorim, M. Bauböck, J. P. Berger, H. Bonnet, W. Brandner, et al., "Detection of the Schwarzschild Precession in the Orbit of the Star S2 near the Galactic Centre Massive Black Hole,"** ***Astronomy & Astrophysics*** **636 (April 2020): L5, https://doi.org/10.1051/0004-6361/202037813.**

4 More specifically, the resolution of a telescope depends on the wavelength it's trying to see and the diameter of its collecting mirror. Radio telescopes are so large—the biggest is the five-hundred-meter aperture spherical radio telescope (FAST) in China—because they're trying to collect large wavelengths of light. But a larger telescope may have a smaller field of view, or some other disadvantage, so it's not always prudent to try to maximize your resolution when planning an observation.

5 The astronomers who discovered this small black hole in its binary orbit with a red giant star named it the Unicorn. At a distance of only 460 pc, this Unicorn black hole may be the closest one to us! **T. Jayasinghe, K. Z. Stanek, Todd A. Thompson, C. S. Kochanek, D. M. Rowan, P. J. Vallely, K. G. Strassmeier, et al., "A Unicorn in Monoceros: The 3 $M_{\odot}$ Dark Companion to the Bright, Nearby Red Giant V723 Mon Is a Non-interacting, Mass-Gap Black Hole Candidate,"** ***Monthly Notices of the Royal Astronomical Society*** **504, no. 2 (June 2021): 2577–602, https://doi.org/10.1093/mnras/stab907.**

6 The Event Horizon Telescope Team collected nearly ten petabytes (that's ten million gigabytes) of data! That information had to be stored on physical data drives because transferring from their remote observing sites over the internet would be horrendously slow. The literal pile of drives then had to be transported to processing centers in Germany and the US. To read more about the data processing and see an awesome picture of Katie Bouman air hugging the ton of data, see **Ryan Whitwam, "It Took Half a Ton of Hard Drives to Store the Black Hole Image Data," ExtremeTech, April 11,**

2019, https://www.extremetech.com/extreme/289423-it-took-half-a-ton-of-hard-drives-to-store-eht-black-hole-image-data.

7 These supermassive black holes in dwarf galaxies are likely offset due to previous collisions with other galaxies, which would help explain how a dwarf got such a massive black hole in the first place. **Phil Plait, "Dwarf Galaxies Have Supermassive Black Holes, Too... and Some Are Off-Center!," SYFY Wire, January 6, 2020, https://www.syfy.com/syfy-wire/dwarf-galaxies-have-supermassive-black-holes-too-and-some-are-off-center.**

8 Quasars are a class of active galactic nuclei (i.e., big energetic black holes) with powerful jets of shining matter streaming away from the center. The word is a shortening for "quasi-stellar" radio source because astronomers thought they were stars when they first observed them in the mid-twentieth century. When the jets are pointed right at us, blasting us with their light, the AGN is called a blazar.

CHAPTER TEN: AFTERLIFE

1 Before we *Homo sapiens* became the dominant human species, there were several early human species living simultaneously. They even interbred. For a fun, interactive human evolutionary tree, see **"Human Family Tree," *What Does It Mean to Be Human?*, Smithsonian National Museum of Natural History, December 9, 2020, https://humanorigins.si.edu/evidence/human-family-tree.**

2 **Bridget Alex, "How We Know Ancient Humans Believed in the Afterlife," *Discover*, October 5, 2018, https://www.discovermagazine.com/planet-earth/how-we-know-ancient-humans-believed-in-the-afterlife.**

3 The exact rules of this game aren't known, but it seems to have been very popular in Mesoamerica because hundreds of courts with standardized dimensions have been found throughout the region. From what we do know (much of it from reports written by invading Spaniards), the game appears to be a mix of soccer and basketball. Two teams of five or so players compete to bounce balls through hoops mounted high on walls, but they can't use their hands or feet.

4 In 2019, Sagittarius A* suddenly flared to one hundred times its normal brightness in the infrared. Astronomers think the flare was likely due to a sudden infall of material. **Susanna Kohler, "Flares from the Milky Way's Supermassive Black Hole," AAS Nova, April 7, 2021, https://aasnova.org/2021/04/07/flares-from-the-milky-ways-supermassive-black-hole/.**

CHAPTER ELEVEN: CONSTELLATIONS

1 Ancient Greek texts say that Andromeda is from Aethiopia, which used to be a general term for the land in Africa south of Egypt. That does cover the

modern-day location of Ethiopia on the eastern coast of the continent by the Red Sea, but some scholars interpret the myth differently. They say that Andromeda was chained to a rock off the coast of Israel. So it's impossible to say where the myth supposedly took place.

2 Medusa was turned into a hideous monster by the goddess Athena after she caught Medusa and the god Poseidon cavorting in her temple. Some interpretations claim that Medusa seduced Poseidon and therefore deserved the punishment. Others claim that Poseidon forced himself on Medusa and then stood by while she bore the brunt of Athena's wrath.

CHAPTER TWELVE: CRUSH

1 Astronomers who study the motion of stars around galaxies pay attention to what we call the galaxy's gravitational potential. This is essentially an equation that describes how matter is distributed throughout the galaxy. Older, more elliptical galaxies tend to have prolate or triaxial potentials because they're more spherical.

2 It's important to consider rest frames in physics because they tell you the point of reference for all motion. The Milky Way's rest frame is centered on its center of gravity, near Sagittarius A*.

3 Acceleration is the change in an object's velocity. In physics, "jerk" is the change in an object's acceleration over time. As the Milky Way and Andromeda approach each other, their acceleration due to the mutual pull of gravity will increase, giving them a positive *jerk*.

CHAPTER THIRTEEN: DEATH

1 Hawking radiation, named after Stephen Hawking (with whom I was honored to share a birthday, along with Elvis and David Bowie), has never been observed. It is a theoretical way for black holes to dissipate their energy. Pairs of particles form on the boundary between a black hole and the vacuum of space, except the particles are just as likely to be on the outside of the black hole as the inside. The particles on the outside escape and carry a tiny bit of the black hole's energy with them.

2 The Large Hadron Collider is a giant circular tunnel built underground in Switzerland. Particles rush around the 16.6-mile circumference, building up speed before they crash into each other with enough energy to produce other, more exotic particles.

CHAPTER FOURTEEN: DOOMSDAY

1 To hear how I believe the nine worlds of Norse mythology line up with the planets in our solar system, listen to the "Norse Cosmology" episode of the *Spirits* podcast. **Amanda McLoughlin and Julia Schifini, "Norse Cosmology," August 12, 2020, in *Spirits*, produced by Julia Schifini, podcast, 49:10, https://spiritspodcast.com/episodes/norse-cosmology.**

2 A few of the planets may line up once every few decades, but it would be nearly impossible for all eight (or nine if you count Pluto) of the planets to form a solar system–wide syzygy. The last time all of the planets were even in the same vague region of the sky was over a thousand years ago. But even if the planets did align, their combined gravitational pull would barely be noticeable to us, and it certainly wouldn't be enough to bring about the end of the world!

CHAPTER FIFTEEN: SECRETS

1 Two teams using supernovae as standard candles independently discovered that the universe's expansion is accelerating. Those teams were the Supernova Cosmology Project led by Saul Perlmutter in California and the High-Z Supernova Search Team led by Brian Schmidt in Massachusetts.

2 The first planet outside of our solar system was discovered in 1992 around a pulsar using the radial velocity method. Three years later, a planet was discovered for the first time around a sun-like star. The discovery of that planet earned Michel Mayor and Didier Queloz the Nobel Prize in Physics in 2019.

3 The transit method was first used to discover an exoplanet in 1999 in a study led by then-graduate student Dave Charbonneau. This was a very big deal that opened the door for exoplanets to *boom* as a subfield of astronomy. Sixteen years later, Dave led my senior thesis seminar at Harvard and was nice enough to pose in a very goofy picture when I turned in my thesis.

4 No astronomer would ever want to hide it if they discovered aliens because it would mean missing out on a guaranteed Nobel Prize. But even if we wanted to keep it a secret, there are protocols in place that say we're obligated to share the information. There is no official government-mandated post-detection protocol, but many organizations have their own. A popular one was published by the Swedish International Academy of Astronautics in 1989: **"Declaration of Principles Concerning Activities Following the Detection of Extraterrestrial Intelligence," International Academy of Astronautics, 1989, https://iaaspace.org/wp-content/uploads/iaa/Scientific%20Activity/setideclaration.pdf.** SETI and NASA have both drafted their own protocols influenced by the IAA's.

5 NASA refused to change the name of the James Webb Space Telescope, even after

more than one thousand astronomers petitioned to change the name. It should be noted that the name wasn't chosen through a typical formal process, and it is not uncommon for telescopes' names to change (e.g., WFIRST changing to the Nancy Grace Roman Space Telescope, or LSST changing to the Vera C. Rubin Observatory). **Nell Greenfieldboyce, "Shadowed by Controversy, NASA Won't Rename Its New Space Telescope," NPR, September 30, 2021, https://www.npr.org/2021/09/30/1041707730/shadowed-by-controversy-nasa-wont-rename-new-space-telescope.**

6 Dr. Jill Tarter has been a champion of SETI research for decades, even when she wasn't serving as the chair of the SETI Institute. She inspired the character Ellie Arroway in Carl Sagan's novel, *Contact*, which was adapted into a movie starring Jodie Foster.

For further reference material
from the author, visit moiyamctier.com.

INDEX

Abell 85 (ACO 85), 131, 133
"absolute zero," 28
accretion disks, 116–17, 123, 129–30
active galactic nuclei (AGN), 117
afterlife myths, 136–43
age of Milky Way, 33–36
agriculture and weather forecasting, 7
air pollution, 44, 212, 215
alien life. *See* extraterrestrial life
Almagest, The (Ptolemy), 147–48
Alpheratz, 147
Andromeda Galaxy, 43, 44, 144–49, 151–68
  Big Freeze and, 179
  Big Rip and, 177
  classification, 147–49
  collision with the Milky Way, 160–68, 171
  distance estimate, 152
  globular clusters, 144, 156–57, 162
  luminosity estimates, 152–54
  mass estimates, 154–55
  mythology and naming, 144–47, 230–31*n*
  nearby and satellite galaxies, 47, 156–60
  observational history, 147, 151–54
  Trin and, 47, 162–63
angular momentum, 70, 80–81, 123–24, 135
anthropocentricism, 93, 207
antineutrinos, 199–200
Antlia, 60
apocalypse, 191–94
arcseconds, 53
Aristotle, 81–82
Assyrians, 14, 148, 192
asterisms, 147–49
astrology, 88, 227*n*
astronomers
  mythology and names, 16–17
  number of, 9
astronomical units (AU), 77, 126
Athena, 146, 231*n*
Avempace (Ibn Bajja), 82
axions, 200

Babcock, Horace, 197
Babylonians, 97
barred spiral galaxies, 73–74, 152
baryonic acoustic oscillations (BAO), 61–63
baryonic matter, 79–80, 125, 173, 181, 197
beryllium, 25, 109
Betelgeuse, 153
Big Bang, 19–22, 28–29, 36, 100, 182, 198
Big Bang nucleosynthesis, 21, 108
Big Bounce, 172, 183–84, 190
Big Crunch, 172, 181–83
Big Freeze, 172, 178–81, 187

Big Rip, 172, 176–78
Big Slurp, 172, 184–87
biosignatures, 210–11
bipedalism, 93, 121
bird migration, 14, 16
Bird's Path, 14, 16
black holes, 113, 116–35, 181, 198. *See also* Sagittarius A*
  Andromeda–Milky Way collision, 166–67
  mass of, 124, 126–27, 128, 213
  observational evidence, 125–32, 174
  properties and structure, 121, 123–24
  use of term, 124–25, 182
Blandford, Roger, 129–30
Blandford-Znajek process, 129–30
blue giants. *See* O-type stars
blueshift, 63–64
body. *See* structure
Boltzmann, Ludwig, 221*n*
Boltzmann Brains, 2–3, 221*n*
Book of Revelation, 191, 192
Book of the Dead, 139
bosons, 185, 186–87
Brahma, 38–39
brown dwarfs, 161, 197, 221–22*n*
burial practices, 137, 139, 170

Cannon, Annie Jump, 113–14
carbon, 25, 33, 36, 58, 108, 109, 112, 228*n*
Carina Dwarf Galaxy, 59
Carina–Sagittarius Arm, 72–73
Cassiopeia, 145, 157–58
cell phones, 11–12
Cepheid variable stars, 56–57, 83–84
Cepheus, 145–46
Cetus, 145–46
Chandrasekhar limit, 58–59
Charbonneau, Dave, 232*n*
charge and black holes, 123
Chinese, ancient, 148–49
chirp, 60
Christianity, 39, 190–93
circumgalactic halo, 78, 161
Clarke, Arthur C., 95
classifications
  of constellations, 147–49
  of dark matter, 199–201
  of galaxies, 83–85
  of stars, 110–11, 113–14
"closed" universe, 176
C-N-O cycle, 228*n*
Coma Cluster, 197
conduction, 107
consciousness, 3
conservation of angular momentum, 70, 80–81
constellations. *See also* Andromeda
  classification of, 147–49
Contact (film), 233*n*
convection, 107–8, 109, 111
cosmic distance ladder, 52–55, *54*, 60, 61, 63
cosmic inflation, 21–22, 173–74
cosmic microwave background (CMB), xi, 22–24, 28
cosmic rays, 198
cosmic redshift, 63–64
cosmological constant, 202–3
"coupled fields," 185
creation myths, 37–41
"critical density," 79–80, 173–76
Curtis, Heber, 83, 129
curvature, 174

dark energy, 79, 97–98, 173, 175–76, 201–3
  Big Freeze and, 172, 178–81
  Big Rip and, 176–78
  black holes compared with, 125
Dark Energy Survey (DES), 203–4
dark matter, 78–81, 196–202
  black holes compared with, 125

characteristics of, 98, 198–200
history and observational evidence, 196–99
theoretical classifications, 199–201
declination, 92, 227*n*
degenerate matter, 58
density of the universe, 173–76, 177–78, 187
density wave theory, 72–74
deuterium, 222
disks, 68–74, 82. *See also* accretion disks
"distance ladder," 52–55, *54*, 60, 61, 63
doomsday myths, 188–94
Doppler effect, 63, 129
Draco, 60
Drake, Frank, 212–13
dung beetles, 222*n*
dwarf galaxies, 6, 29–32, 44, 48, 50–51, 85, 130–31, 156, 177
dwarf planets, 29–30
Dyson, Frank, 82
Dyson, Freeman, 226*n*
Dyson sphere, 212, 226*n*

Earth Diver myths, 40–41
Eddas, 39–40
Eddington, Arthur, 112
effective radius, 69
Egyptians, ancient, 138–39, 148, 222*n*
Einstein, Albert, 173–74, 183–84, 187, 201, 202
electromagnetic radiation, 109, 111–12, 127–28
electromagnetic spectrum, 22–23, 64
electromagnetism, 101, 185, 198
electron degeneracy pressure, 58
electrons, 185, 228*n*
elliptical galaxies, 84–85, 104, 155, 166–67
End, The, myths about, 188–94
endoplanets, 205
entropy, 2
epicycles, 69, 70
Euler's number (number *e*), 68–69
event horizon, 123, 127–29
Event Horizon Telescope (EHT), 128–29, 229*n*
evolution, 1, 137, 209, 224*n*, 230*n*
"exomoons," 226*n*
exoplanets, 205–10
expansion of the universe, 20–24, 26, 29, 43, 61, 63–64, 79, 98, 174–76, 182, 202–3, 232*n*
Big Freeze, 172, 178–81
Big Rip, 172, 176–78
extraterrestrial life, 75, 78, 204–13
humanoid appearance of, 93–94

"false vacuum decay," 186
FAST telescope, 229*n*
Fenrir, 189–90
Field of Reeds (Field of Rushes), 139
Fimbulwinter, 189
force particles, 185
Fornax, 60, 157
fractal cosmology, 60
Friedmann, Alexander, 173–74
Friedmann equations, 174–75
fundamental forces, 101–2, 111
fundamental particles, 184–85

Gaia (spacecraft), 86, 154
Gaia Enceladus (GE), 156–57
galactic archaeology, 33–36, 162
galactic bulge, 74–78, 87
galactic coordinate systems, 91–92
galactic halo, 78–84, 161–62
galaxies. *See also* Andromeda Galaxy; Local Group; Milky Way
classification of, 83–85
death of. *See* ultimate fate of the universe
formation of, 19, 25–26, 29–31
neighborhood, 43–52

(galaxies *cont.*)
size of, 29–31, 68
use of term, 81–82
galaxy clusters, 151–52, 181
galaxy effective radius, 69
galaxy rotation curve, 80–81, 197
Galileo Galilei, 82
general relativity, 174
"generation of stars," 25, 100
gigaanna, 31–32
global positioning technology, 12
globular clusters, 78, 83–84, 103–4, 144, 156–57, 162
gluons, 185
GN-z11, 64
gold, 223*n*, 228*n*
Goldilocks zone, 207–10
gravitational potential, 166, 231*n*
gravity, 6, 24, 26, 31, 56, 61–63, 68, 70, 73, 75, 79, 101, 126, 132, 151–52, 200–201
Greeks, ancient, 51, 81, 139, 144–45, 230–31*n*
G-type stars (yellow dwarfs), 108–11
gyrochronology, 35

habitability question, 78, 207–10
half-light radius, 69
halo, 78–84, 161–62
Hawking, Stephen, 231*n*
Hawking radiation, 181, 231*n*
Heat Death of the universe, 178–81
Heavenly Market, 148
heavy metals, 24–25
Heimdall, 189–90, 190
helium, 22, 23, 24, 25, 100
fusion process, 57–58, 106, 108, 109, 111, 112, 114, 228*n*
Hermes, 146
Herschel, Caroline, 82, 226*n*
Herschel, William, 82, 226*n*
Hertzsprung-Russell diagram, 114
higglet, 200
Higgs boson, 186–87
Higgs field, 184, 186–87
Hinduism, 38–39, 139–40
Hipparchus, 152–53
Hodierna, Giovanni Battista, 46–47
Holmberg 15A, 131, 133
hot Jupiters, 4, 207, 222*n*
Hubble, Edwin, 83–85
"Hubble constant," 174–75
Hubble parameter, 174
Hubble Space Telescope, 85–86, 87, 204, 226*n*
Hubble tuning fork, 84–85
human evolution, 1, 137, 209, 230*n*
human life span, 96, 169–70
human race, myth of united, 89
Hunahpu, 140, 142
hydrogen, 22, 23, 24, 47, 77, 100, 175
fusion process, 57–58, 106, 108, 109, 112, 114, 228*n*
hydrostatic equilibrium, 103, 108, 109, 111, 112
hypervelocity stars, 158, 160
hypothesis vs. theory, 72

industrialization, 9
inflationary epoch, 21–22, 173–74
infrared, 22, 23, 126
initial mass function (IMF), 106, 213, 228*n*
intergalactic gas, 43–44, 49, 132–33
International Astronomical Union (IAU), 17, 149
International Dark-Sky Association, 222*n*
interstellar political coalition, 89, 91
iron, 25, 112
irregular dwarf galaxies, 50, 85
isochrones fitting, 34, 35

James Webb Space Telescope (JWST), 211–12, 232–33*n*

jansky, 118
Jansky, Karl, 118–19
Jeans, James, 196
jellyfish galaxies, 133, 135
"jerk," 165, 231*n*
Jesus, 191, 192
John of Patmos, 190–91, 192
Jörmungandr, 189
JO201 ("Jo"), 133, 135, 136
Jukara, 51
Jupiter, 97, 195, 211

Kelvin, William Thomson, 1st Baron, 196, 197, 198
Kepler, Johannes, 126–27
Kepler orbits, 127
Kepler space telescope, 86
Khoisans, 14
K'iche', 140, 142
kiloparsecs, 49, 68, 69, 91, 210
Kinich Ahau, 140
Kroupa, Pavel, 106, 228*n*

Lambda-CDM model, 202
LAMOST-HVS1, 160
Large Hadron Collider, 199, 231*n*
Large Magellanic Cloud ("Larry"), 31, 37, 44, 48–51, 56–57, 82, 97, 115
Leavitt, Henrietta, 56–57, 83, 225*n*
Lemaître, Georges, 183–84
lenticulars, 85
light pollution, 9, 11, 44, 212, 222*n*
Lindu, 14, 16
lithium, 22, 23, 35
Local Group, 43–52, 59–61, 143, 152. *See also* Andromeda Galaxy
  Gaia Enceladus (GE), 156–57
  Large Magellanic Cloud ("Larry"), 31, 37, 44, 48–51, 56–57, 82, 97, 115
  Small Magellanic Cloud ("Sammy"), 44, 46, 49–51
  Triangulum ("Trin"), 44, 46–48, 111–12, 162–63
Loki, 189–90
Lowell, Percival, 205
low-mass O-type stars, 113
low-mass stars, 105–6
luminosity, 55–59, 112, 152–54

Mach'acuay, 149
Magellan, Ferdinand, 51
"Magellanic spirals," 48–49, 225*n*
magnitude, 152–53
major mergers, 164
Maori, 50–51
Mars, 97, 204–5
mass
  of Andromeda, 154–55
  of black holes, 124, 126–27, 128, 213
  of stars, 34, 58, 105–6, 112, 228*n*
mass-extinction events, 109–10, 224*n*
massive astronomical compact halo objects (MACHOs), 197–98
"massive" stars, 56, 106, 108, 110–12, 116
matter particles, 185
Mayans, 140, 142, 193
mayflies, 32, 223–24*n*
Mayor, Michel, 232*n*
MCG+01-02-015.7 ("Macy"), 62
Medusa, 146, 231*n*
MeerKAT, 87, 128
Mercury, 97, 124
messenger stars, 157–58, 160
Messier, Charles, 47
Messier objects, 46–47
Messier 31 (M31). *See* Andromeda
Messier 32 (M32), 163–64
Messier 33 (M33). *See* Triangulum
Messier 87 (M87), 129–30, 131
microwave, 22, 23

Milgrom, Mordehai, 201
Milky Way
  age of, 33–36
  creation myths, 37–41
  death of. *See* ultimate fate of the universe
  distance measurement, 51–65
  etymology of name, 13–17
  formation of, 4, 6–7, 18–36. *See also* Big Bang
  introduction to, 1–12
  relationship with Andromeda, 151–68
  structure. *See* structure
minor mergers, 164, 171
mobile phones, 11–12
modern science-fiction myths, 88–95
modified Newtonian dynamics (MOND), 201
moon, ix, 35, 97, 149, 224*n*
Mount Vesuvius, 192
M-type stars (red dwarfs), 106–9, 179, 182, 227*n*
mummification, 139
muons, 185
Muspelheim, 39–40
mythology (myths), xi–xii, 32, 37–38, 50–51, 88–89
  afterlife, 136–43
  Andromeda, 144–46, 230–31*n*
  doomsday, 188–94
  Milky Way creation myths, 37–41
  naming of Milky Way, 13–17
  science fiction, 88–95

Nancy Grace Roman Space Telescope, 203–4, 233*n*
Native Americans, 40–41
nebulae, 94–95
Nereids, 145
neutrinos, 113, 185, 199–200, 228*n*
neutrons, 21, 36, 101–2, 113
neutron stars, 112–13, 128, 223*n*, 228*n*
Newton, Isaac, 200–201
Next Generation Space Telescope, 211–12
NGC 1052-DF2, 201
NGC 1052-DF4, 201
NGC 1300, 73–74
nitrogen, 25, 109, 112
Nobel Prize in Physics, 121, 232*n*
Norma Arm, 72
Norse mythology, 39–40, 51, 189–90, 232*n*
nuclear fusion, 101, 102–3, 108

observable universe, 64–66, 68
Odin, 190
Oort, Jan, 196–97
"open clusters," 35, 103–4
"open" universe, 176
Orion-Cygnus, 72
Orion Nebula, 94
Osiris, 139, 222*n*
O-type stars (blue giants), 111, 112–13, 114, 182, 227*n*
overview effect, xii

parallax, 52–55, 59
parsecs, 53, 68, 82, 226*n*
Perlmutter, Saul, 232*n*
Perseus, 146
Perseus Arm, 72–73
Phoenix Cluster, 59, 157
photons, 185, 199, 200
photosynthesis, 77, 210
Pickering, Edward, 225*n*
Pinwheel Galaxy. *See* Triangulum
planet, origin of term, 97
Poincaré, Henri, 196
Polaris, 153, 222*n*
Population III stars, 24–25
Poseidon, 145–46, 231*n*
"primordial atom," 184

primordial black holes, 198
primordial dwarfs, 30–31
protogalaxies, 26, 75, 80
protons, 21, 36, 101–2, 113, 123, 128, 179, 186, 198
proton decay, 179, 181
proton-proton (p-p) chain, 228*n*
Ptolemy, 147–48
pulsation, 56
Purple Forbidden, 148

quadrants, 91, 92
quantum fields, 185–86, 202
"quantum tunneling," 187
quarks, 102, 185, 186, 197, 199
quasars, 131, 230*n*
Queloz, Didier, 232*n*
quenching, 104–5, 135
quintessence, 202–3

Ra, 139
radial velocity, 206, 232*n*
radiation, 107, 173, 208, 209. *See also* electromagnetic radiation; Hawking radiation
radiation pressure, 109, 111–12
radio astronomy, 118–19
radio waves, 11–12
Ragnarok, 189–90
"ram pressure," 133
Reagan, Natalia, 229*n*
red dwarfs, 106–9, 179, 182, 227*n*
"red giants," 109–10, 228*n*
redshift, 63–64
reincarnation, 139–40
rest frames, 231*n*
Rigel, 153
right ascension (RA), 92, 227*n*
Romans, ancient, 192
rotating black holes, 123, 129–30
rotation curves, 80–81, 197
RR Lyrae, 56–57
Rubin, Vera, 80–81, 197
Rutherford, Ernest, 185

Sagan, Carl, 11, 233*n*
Sagittarius (constellation), 44, 83, 118–19
Sagittarius A* ("Sarge"), 117–21, 135
  central black hole, 75, 119, 124, 126–27, 132–33, 143, 167
  history and observation, 87, 118–20, 125–32
  naming of, 118, 119
  properties and structure, 121, 123–24, 126
Sagittarius Arm, 72–73
Sagittarius Stream, 44, 224–25*n*
Sakharov, Andrei, 179, 181
Salpeter, Edwin, 228*n*
Salpeter IMF, 106
Saturn, 97, 229*n*
Schiaparelli, Giovanni, 204–5
Schmidt, Brian, 232*n*
Schwarzschild precession, 229*n*
Schwarzschild radius, 127
science fiction myths, 88–95
scientific method, 64–65, 72
Sculptor Galaxy, 59
Scutum–Centaurus Arm, 72–73
SDSS J090745.0+024507, 158, 160
Search for Extraterrestrial Intelligence (SETI), 212–13
second law of thermodynamics, 2
shape
  of constellations, 147–48
  of galaxies, 84–85
  of the universe, 40, 174, 175–76
Shapley, Harlow, 83
Shapley-Curtis Debate, 83
Shiva, 39
silicon, 112
Sirius, 153, 222*n*
Sloan Digital Sky Survey (SDSS), 62–63, 225*n*

Small Magellanic Cloud ("Sammy"), 44, 46, 49–51
solar axions, 200
solar storms, 8
"spaghettification," 132
speed of light, 6, 22, 113, 126–27, 203
spiral arms, 68, 72–73
spiral galaxies, 48–49, 50, 84–85, 155, 166–67
spiral nebulae, 83
spurs, 72–73
Sputnik, 95
standard candles, 55–59, 75
standard model, 101, 199–200
"standard sirens," 60
stars
  age of, 33–36
  death of, 98, 100, 104, 105–6, 110
  formation of. *See* star formation
  Hipparchus and star catalog, 152–53
  magnitude, 152–53
  mass of, 105–6, 112, 228*n*
  use of term, 96–97
star formation, 4, 6, 98, 100–105, 114
  Andromeda and, 163, 166
  in batches, 103–4, 227*n*
  first stars, 23–25, 29, 100
  quenching, 104–5
  rates, 49, 104–5
Star Trek, 89–90, 91, 92, 95
Star Wars, 92
stellar age determination, 33–36
stellar archaeology, 33–36, 162
stellar black holes, 117, 123
stellar classification, 110–11, 113–14
stellar formation. *See* star formation
stellar halo, 78
stellar isochrones, 34, 35
stellar mass, 105–6, 112, 228*n*
stellar parallax, 52–55, *54,* 59
stellar population, 24–25
stellar pulsation, 56
Stöffler, Johannes, 192–93
Straw Thief's Way, 13–14
strong charge-parity (CP) problem, 199–200
strong nuclear force, 102–3, 112, 177, 185, 198
structure, 67–87
  disk, 68–74, 82
  galactic bulge, 74–78
  halo, 78–84
S-type stars, 126–27
Sufi, Abd al-Rahman al-, 144
sun, ix, 82–83, 97
  orbits, 72–73
  red giant phase, 109–10, 228*n*
  stellar age for hosting life, 36, 224*n*
  white dwarf phase, 57–58, 108
super-Chandras, 58
supernova explosions, 75, 77, 209
"supersonic," 133
Supreme Palace, 148
synchrotron radiation, 127–28

Tarter, Jill, 212–13, 233*n*
taus, 228*n*
technosignatures, 212–13
telescopes, 8, 127–28, 229*n*. *See also specific telescopes*
temperature, 20–21, 26, 28
thin disk, 68–74
timekeeping, 8, 11
transit photometry, 206, 211, 232*n*
transit spectroscopy, 211
transmission spectroscopy, 211
Triangulum ("Trin"), 44, 46–48, 111–12, 162–63

type II supernovae, 112
type 1a supernovae, 57–58, 59, 75, 77

ultimate fate of the universe, 169–87
  Big Bounce, 172, 183–84
  Big Crunch, 172, 181–83
  Big Freeze, 172, 178–81
  Big Rip, 172, 176–78
  Big Slurp, 172, 184–87
universe, the
  cooling down, 20–21, 26, 28, 223*n*
  expansion of the. *See* expansion of the universe
  formation of, 4, 6–7, 18–36. *See also* Big Bang
  as opaque, 22, 23
  shape of the, 40, 175–76
  as transparent, 22
  ultimate fate of. *See* ultimate fate of the universe
Uranus, 97

Vahagn, 14
Valhalla, 140, 190
variable stars, 56–57
Vega, 153
Venus, 97, 210
Vidar, 190
Virgo Cluster, 60–61, 129
Virgo Supercluster, 60–61, 85, 129
Vishnu, 39, 192

wandering stars, 96–97
warm-hot intergalactic medium, 43–44, 49
water, 40–41, 207–8
Way of the Birds, 14, 16
weakly interacting massive particles (WIMPs), 198–99
weak nuclear force, 101–2, 185, 198
weather forecasting, 7
Wheeler, John, 182
Whirlpool Galaxy, 73
White, Frank, xii
white dwarfs, 57–58, 77, 108, 110
Williams, Sarah, ix
Wright, Thomas, 82

Xbalanque, 140, 142
XENON1T, 200
Xibalba, 140, 142
X-rays, 12, 117

Yacana, 149
yellow dwarfs, 108–11
Yggdrasil, 40, 190
Ymir, 39–40

Znajek, Roman, 129–30
Zwicky, Fritz, 197

YOUR BOOK CLUB RESOURCE

READING GROUP GUIDE FOR
THE MILKY WAY

DISCUSSION QUESTIONS

1. McTier's approach to explaining the science of the galaxy we live in is unique. Why do you think she chose to write this project from the perspective of the Milky Way? How might your reading experience have differed if the book were in McTier's own voice?

2. Mythology has played a central role in our relationship with the Milky Way throughout history. Why do you think this galaxy has loomed so large in lore across time and cultures? What might you have thought of the Milky Way, had you spotted it with no prior scientific awareness?

3. The bright lights of cities and towns obscure views of galaxies, stars, and other planets, so we can't as easily see all that our ancestors could. Consider the last time you got an unobscured look at the night sky. What, if anything, do you think we've lost in being unable to access that view regularly?

4. In the Milky Way's estimation, all humans share a relatively small hometown—Earth—yet we're driven apart by inconsequential differences. Does considering humanity from a galaxy's point of view shift your perspective? How might we benefit from thinking of other people, and ourselves in relation to them, from this more macro viewpoint?

5. In the same vein, the Milky Way has a very particular perspective on human life. How does its commentary on everything from the brevity of our lives to the relative magnitude of our failures shift your outlook?

6. Throughout the book, there are references to a variety of telescopes and other tools named for human scientists, and then renamed. What would you like to be named after yourself and why? What is the power of a name?

7. There are references to several underappreciated scientists through history in this book. Who seems to get left out of our popular scientific narratives and why?

8. The Milky Way outlines human scientific progress through history, while also alluding to how much we have yet to learn. How has this narrative shaped your understanding of the daily work of scientists and scientific advancement? Did it dispel any misconceptions you had?

9. In the science fiction chapter, the Milky Way delves into not only what the genre has gotten wrong, but also the benefits of storytelling. What are your favorite science fiction properties? How have they shaped your understanding of space, as well as life here on Earth?

10. In the constellations chapter, the book focuses in on the monikers and stories that various cultures have associated with different collections of stars. What do these pieces of lore reveal about the cultures that developed them?

11. The Milky Way writes, "Sarge stops me from envisioning what I could be because I'm too busy lamenting what I've never been and despairing what I was instead." What is your Sarge? Did the Milky Way's relationship to its central supermassive blackhole encourage you to reconsider your own demons?

12. The Milky Way knows that, despite all the advancements our scientists have made, there are questions we haven't even thought to ask yet. What questions do you think scientists will start asking next? How do you think science and technology will progress over the next hundred years?

ABOUT THE AUTHOR

Dr. Moiya McTier is an astrophysicist, folklorist, and science communicator. After graduating from Harvard as the first person in the school's history to study both astronomy and mythology, McTier earned her PhD in astrophysics at Columbia University, where she was selected as a National Science Foundation research fellow. McTier has consulted with companies like Disney and PBS on their fictional worlds, helped design exhibits for the New York Hall of Science, and given hundreds of talks about science around the globe. You can see McTier as a cohost of PBS's *Fate & Fabled* and hear her on her two podcasts: *Pale Blue Pod* about astronomy and *Exolore* about fictional worldbuilding.